Oxford Studies in Normative Ethics

Oxford Studies in Normative Ethics

Volume 12

Edited by

MARK TIMMONS

OXFORD
UNIVERSITY PRESS

OXFORD
UNIVERSITY PRESS

Great Clarendon Street, Oxford, OX2 6DP,
United Kingdom

Oxford University Press is a department of the University of Oxford.
It furthers the University's objective of excellence in research, scholarship,
and education by publishing worldwide. Oxford is a registered trade mark of
Oxford University Press in the UK and in certain other countries

Published in the United States of America by Oxford University Press
198 Madison Avenue, New York, NY 10016, United States of America

British Library Cataloguing in Publication Data

Data available

Library of Congress Control Number: 2022931795

ISBN 978-0-19-286888-6

DOI: 10.1093/oso/9780192868886.001.0001

Printed and bound in the UK by
TJ Books Limited

Contents

Acknowledgments

Versions of the chapters in this volume were presented at the 2021 Arizona Workshop in Normative Ethics, January 14–16, 2021. I thank the Center for the Philosophy of Freedom and the Department of Philosophy at the University of Arizona for their generous financial support of the workshop. Of course, the views expressed in the volume's chapters do not necessarily reflect the views of the Center or the Department.

I am grateful to Cheshire Calhoun, Josh Glasgow, Robert Johnson, Mark LeBar, Doug Portmore, Mark Van Roojen, and Chelsea Rosenthal for serving as referees for the 2021 program and to two anonymous referees for Oxford University Press who wrote excellent comments on the workshop papers and gave constructive advice to authors. Thanks also to Caleb Dewey who prepared the volume's index, to Betsy Timmons who greatly facilitated conducting the workshop online, and to my editor, Peter Momtchiloff, for his support.

Mark Timmons
Tucson, AZ

List of Contributors

Joseph Bowen is a Postdoctoral Researcher at the Stockholm Centre for the Ethics of War and Peace, Stockholm University. From August 2022, he will be a Lecturer in Philosophy at the University of Leeds.

Adam Cureton is Professor of Philosophy at the University of Tennessee, Knoxville.

Michael Deigan is an Andrew W. Mellon Postdoctoral Fellow in philosophy at Rutgers University.

Robin S. Dillon is Professor Emeritus at Lehigh University.

Jamie Dreier is Judy C. Lewent and Mark L. Shapiro Professor of Philosophy at Brown University.

Emma Duncan is a Doctoral Candidate in Philosophy at the University of California, San Diego.

Jessica J. T. Fischer is a Postdoctoral Fellow at the Justitia Center for Advanced Studies, Goethe University Frankfurt.

Paul Hurley is Edward J. Sexton Professor of Philosophy at Claremont McKenna College.

Sarah McGrath is Professor of Philosophy at Princeton University.

Douglas W. Portmore is Professor of Philosophy in the School of Historical, Philosophical and Religious Studies at Arizona State University.

Mark Schroeder is Professor of Philosophy at the University of Southern California.

Maria Seim is a University Lecturer in the Department of Philosophy, Classics, History of Art and Ideas at the University of Oslo.

Ralph Wedgwood is Professor of Philosophy at the University of Southern California.

Introduction

Mark Timmons

Oxford Studies in Normative Ethics features new work on topics in normative ethical theory. This twelfth volume features chapters on the following topics: the vices of greed and arrogance, harmless wronging, Kantian ethical theory and partialist reasons, moral contractualism, explaining value comparisons in cases of parity, weighing reasons for action, the burdens of trust, attributive silencing, forgiveness, offsetting harm, paternalism and interpreting others, consequentializing moral theories, and the nature of moral worth.

Robin S. Dillon in "Old-Fashioned Vices in Contemporary Crises; or, It Matters How You Value Yourself" argues that vices share the feature of being grounded in distorted self-valuing—an old-fashioned view at the center of the Deadly Sins tradition going back to the fourth century CE. Dillon points out that Kant's conception of vice is similarly committed to this "distorted self-valuing" idea of vice which she proceeds to develop focusing on greed and arrogance. Dillon considers the hoarding of goods during the COVID-19 pandemic and speculates that greed is the likely deep motivation behind such hoarding and that not only does it ignore the needs of others and in some cases even treats others merely as means to selfish ends, but it is also a form of "disordered self-valuing" and thus self-disrespectful. Ways of disrespecting others as well as oneself are central in the vice of arrogance, including what Dillon calls "status arrogance" (looking down on others as being of lesser moral importance than oneself) and "unwarranted claims arrogance" (for example, claiming superior authority, rights, or knowledge that is undeserved). Again, the common core of these two species of arrogance, she argues, is a distorted sense of self-valuing. As Dillon proceeds to argue, greed's disrespect of others and oneself involves both forms of arrogance. The fact that greed has its source in a distorted sense of oneself manifested in types of arrogance is what makes it a self-corrupting deadly vice.

Mark Timmons, *Introduction* In: *Oxford Studies in Normative Ethics, Volume 12*. Edited by: Mark Timmons, Oxford University Press. © Mark Timmons 2022. DOI: 10.1093/oso/9780192868886.003.0001

Cases of harmless wrongdoing pose a challenge to rights theories. These are cases in which one's rights against being harmed are violated, yet the behavior in question harms no one. For instance, in so-called preempted harms cases, one's rights are violated, yet the violation turns out to save one from harm that would have otherwise occurred had the right not been violated. In risk imposition cases, risky behavior puts others in danger of harm, yet no harm comes about; still, rights are violated. As **Joseph Bowen** explains in "Robust Rights and Harmless Wronging," this harmless wronging challenge seems especially powerful against interest theories of rights according to which rights necessarily protect a rights holder's well-being. In dealing with the challenge, Bowen proposes a modally robust "safety condition" according to which in assigning rights one looks beyond what happens in the actual world and takes into account what could easily have happened, thus "for someone to hold a right against us that we do not perform some action, we look to whether our performing that action could easily leave them sufficiently worse off to place us under a duty." In effect what the safety condition does in yielding correct verdicts about harmless wronging cases is to eliminate objectional luck from rights and rights violations. In defending his proposal, Bowen replies to a series of possible objections including that it suffers from obscurity and that it succumbs to counterexamples by both under- and over-generating rights.

Kantian moral theory involves a commitment to the primacy, authority, and universality of impartial reason, and thereby seems too "thin" to capture and explain the many reasons to form, maintain, perfect, respect, and promote loving relationships. In short, Kantian moral theory does not seem compatible with the sort of partiality that loving relationships involve. Rejecting appeals to the duty of beneficence and instrumentalist attempts to reconcile Kantian moral theory with such partiality, **Adam Cureton** in "Solidarity in Kantian Moral Theory" appeals to a largely unappreciated theme in Kant's thinking that the faculty of reason has "substantive and intrinsic interests" that provide a basis for explaining the kinds of partialist reasons involved in loving relationships. Besides such formal interests as eliminating contradictions and systematic unification, Cureton argues that reason also includes substantive intrinsic interests including a rational interest in relationships of solidarity realized in social commitments, love, as well as trust in both. The rational interest in solidarity can then help determine which candidate moral laws are rationally justifiable to everyone via Kant's Categorical Imperative, thereby grounding genuine reasons and requirements that capture and explain what we commonly take to be reasons and requirements of partiality.

Contractualist approaches to moral theory following the lead of T. M. Scanlon's version impose an "individualist constraint" on the justification of moral principles that represents its anti-utilitarian core. In effect, this constraint bars individuals from appealing to aggregate claims of individuals as members of some group in the process of justifying moral principles; rather, reasons in such contexts need to be "personal reasons" that appeal solely to the well-being or status of the individual lodging the complaint. This protects against imposing burdens on a few to benefit the many. One issue for contractualists, and the topic of Jessica J. T. Fischer's "The Individualist Objection," is how to evaluate complaints individuals may have toward principles of risk imposition. On so-called ex ante versions of the view, evaluation of complaints in risk situations is understood in terms of the prospects that individuals have based on the probabilities of harm that a principle imposes on individuals. This version is often taken to be superior to an "ex post" version that would consider the complaints individuals would have were they to be negatively affected by the acceptance of a principle. The ex ante version is the particular focus of Fischer's chapter. An ex ante version faces the problem of "ex ante rules," according to which there are cases in which the view yields intuitively wrong verdicts (see Fischer's chapter for examples). Fischer considers but rejects two attempts to overcome the problem, revealing that ex ante contractualism "invariably permits the imposition of large burdens on some, to secure small benefits on others, so long as very low probabilities of harm are imposed on the former, and high probabilities of benefitting on the latter." Fisher's diagnosis of this problem is that ex ante contractualism violates the individualist constraint by allowing complaints against principles whose justification essentially appeals to a property shared with others. If fully satisfying the individualist constraint is a core tenet of contractualism, then ironically the ex ante view fails to be a version of contractualism.

According to the Principle of Personal Good, PPG (due to John Broome, 1995), "If one option is better for someone and at least as good for everyone, then it is Better" (where "Better" refers to ways the world could be). This principle in effect takes all the good in a distribution to be located in separate persons; it is a kind of "separability principle" that explains why one distribution is better than another based on how good each distribution is for each affected person. Separability principles provide guidance and explain value comparisons whenever the value orderings are complete. But consider cases of so-called parity in which one outcome is neither better nor worse than nor equally good as another outcome; they are apparently incomparable. In "Blessed Lives, Bright Prospects, Incomplete Orderings" **Jamie Dreier** explains that PPG and similar principles are not helpful in parity cases and proceeds to consider whether amended separability principles are helpful in such cases, arguing that the verdicts they yield are implausible. Dreier then turns to a different sort of principle, so-called prospectivist principles, and finds that although they work when incomparability is due to ignorance or indeterminacy, they fare less well when it is due to parity.

In situations in which one is confronted with alternative courses of action and there are reasons of kind R that favor one course and reasons of kind R* that favor taking an alternative course, is it possible to somehow aggregate these reasons to reach an all-things-considered (ATC) verdict about which course of action one ought to choose? If so, then how? In "The Reasons Aggregation Theorem" **Ralph Wedgwood** answers this question, proposing an account of weighing reasons according to which they are subject to aggregation. Wedgwood's proposal relies on various premises. First, reasons for or against action are grounded in values, that is, in facts in virtue of which actions are respectively good or bad in some respect and to some degree. Second, each kind of goodness is reducible to the corresponding kind of betterness (a version of Broome's (1999) reduction of goodness to betterness). Combining these premises yields the principle that a reason for a course of action A is a fact in virtue of which A is in a relevant respect better than some relevant alternatives. (Similarly, mutatis mutandis for reasons against.) Together with further premises about cardinality and how specific-reason-giving values are related to the comparison of alternatives bearing on how much reason ATC there is for the alternatives, Wedgwood arrives at an account of the aggregation of reasons that is additive, analogous to John Harsanyi's Social Aggregation Theorem in social choice theory.

Sometimes the prospect of being a trustee is unwelcome. But what is it about the nature of trust that best explains unwelcome trust?—the question **Emma Duncan** in "The Normative Burdens of Trust" addresses. The task involved in answering this question is to "locate" the burden of trust in a way that properly explains some of trust's essential characteristics, including the fact that a sense of betrayal (and not just resentment) is a fitting reaction to having broken another's trust and that trust is a particular species of reliance. With these characteristics at hand, Duncan considers extant attempts to explain unwelcome trust including: (*i*) views that locate the burden in the demandingness of meeting the trustor's expectations, (*ii*) views that locate the burden in not wanting to be susceptible to such reactive attitudes as feelings of betrayal, and (*iii*) views that locate it in certain unwelcome aspects of the trustee–trustor relationship. After explaining why these proposals are problematic, Duncan proceeds to advance a care-based account of trust according to which a normative expectation internal to trust is that the trustee is invited to adopt a particular orientation of care toward the trustor. The contours of this orientation, according to Duncan, serve to plausibly explain the normative burdens of trust and why trust can be unwelcome.

In his chapter "Attributive Silencing" **Mark Schroeder** addresses cases of communicative acts in which one's communication is falsely dismissed as not attributable to one's person but rather written off by one's interlocutors owing to circumstantial obstacles that interfere with correct attribution. Such silencing, besides being hurtful, is morally problematic because it fails to treat one as a person with authority over what is attributable to *them*. Schroeder's chapter raises the question of how easily and thus perva-sive such silencing is, particularly for members of oppressed groups, the answer to which raises the issue of explaining *how* this phenomenon is possible. Schroeder pursues this issue by offering a theory of the conditions of attributability which, he notes, requires providing a theory of persons. According to his view, as a person one is constituted by the best interpret-ation of what does and does not count as truly oneself. Such interpretations are "biased toward the good" in the sense that there is a role for charity in how we interpret others. (A view Schroeder defends in his 2019 essay and discussed by Sarah McGrath in her contribution to this volume.) As he proceeds to explain, on his view attributive silencing is likely common owing to widespread and systematic distortion of values in gendered social worlds where, for example, women's attempts to refuse sex are easily silenced.

One debate over the nature of forgiveness is whether third-party forgive-ness is possible. For example, does one ever have the standing to forgive a wrongdoer for what they did to a loved one? Debates over this issue are often framed within a standard view of the nature of forgiveness according to which genuine forgiveness is a matter of overcoming resentment for the right reasons—reasons that make forgiveness fitting (rather than being prudent or otherwise morally motivated), a view compatible with third-party forgiveness. **Maria Seim** in "The Standing to Forgive" argues that (*i*) a version of this standard account that makes forgiveness conditional on the wrongdoer repenting is more plausible than unconditional versions that do not impose this requirement, but that (*ii*) the conditional version is not able to accommodate how forgiveness is under a person's voluntary control—that forgiveness is something that one can choose to do at will. Importantly, this preserves the thought that forgiveness can be something a person ought to do without it being a duty. Thus, even if a victim has reasons that make forgiving fitting, it is still up to the victim to decide whether to forgive. On the conditional version of the standard view, when a wrongdoer has repented, the victim *must* forgive on pain of irrationality or immorality. The challenge, then, according to Seim, is to explain how forgiveness is subject to rational ought-claims and yet is something one can decide to do or not to do at will. Seim proceeds to argue that to accommodate the kind of voluntary control one exercises in cases of forgiveness the standard account must be rejected in favor of a view according to which (roughly) to forgive is to decide to "absorb the cost" of the wrongdoing and to decide not to hold it against the victim. This involves an attitudinal change in the victim but does not require overcoming resentment; one only commits to relinquishing *expressing* resentment. In Seim's view, repentance is nevertheless a condition of forgiveness because it is the right sort of reason for absorbing the cost of wrongdoing. Her view implies that only a victim has the standing to forgive because only a victim has the standing to decide to "absorb the cost" of having been wronged.

One standard deontological constraint is against doing harm. W. D. Ross, for instance, includes a duty of non-maleficence among the basic pro tanto duties. However, as **Michael Deigan** argues in "Offsetting Harm," this sort of constraint cannot accommodate cases of offsetting harm whereby it is morally permissible to "contribute to a harm in a positive way—in a way that worsens it—so long as one offsets this contribution by doing something that leaves its *net* effect of that harm neutral or negative." This implies that doing something to increase harm, such as CO_2 emission, is not itself pro tanto wrong, while contributing to it in a net positive way is. If this is right, then the standard deontological constraint against harm should be replaced by one the prohibits non-offsetting harms. Defending this proposal requires explaining what *makes* offsetting harms permissible. Rejecting the view that offsetting harm is permissible because harm is prevented (an account that appeals to the standard constraint), Deigan proceeds to offer alternative ways of formulating a non-offsetting harm constraint explaining how they apply to cases of CO_2 emission and meat consumption—cases in which one does harm, but the harm is offset and so not pro tanto wrong. He concludes by considering some objections, including whether there is a plausible rationale for his proposed harm constraint that explains why offsetting harms are permissible.

P. F. Strawson famously distinguishes two stances one might take toward others: a participant stance that engages with the other as a person and an objective stance that views the other (at least on some occasions) as a thing subject to causal forces. Sometimes, it seems proper to move from the participant to the objective stance in one's interpretation of others as a way of not diminishing them as persons. In "Please Keep Your Charity out of My Agency: Paternalism and the Participant Stance" **Sarah McGrath** raises questions about the norms that govern such transitions from participant to objective stance toward others. That is, when should one interpret another's behavior merely causally and thus treat them as a thing? McGrath considers Mark Schroeder's (2019) proposal that we should deploy a principle of charity according to which we switch to an objective stance when doing so makes our interlocutor's contribution to the world "greater, better, or more significant" than it would be if we were to attribute their behavior to them. Schroeder's rationale for his principle of charity is that by deploying it one is more likely to interpret someone's behavior accurately. Despite its various advantages, McGrath questions this principle of charity, arguing that interpreting someone charitably can be objectionably paternalistic— substituting your interpretation of what they mean (in a conversation) with what they meant. In place of a principle of charity, McGrath defends a "default to trust" principle of interpretation: "absent good reason to doubt that the speaker is sincere and accurate, believe her to be sincere and accurate," which gives one a sufficient reason to take (as default) the participant stance toward others and avoids treating them paternalistically.

The project of consequentializing is to show that for any plausible target non-consequentialist moral theory (for example, versions of deontology and virtue ethics) there is a version of consequentialism that is the target's deontic equivalent in the sense that the deontic implications of the two are the same. Thus, any plausible moral theory (more precisely any plausible theory of deontic evaluation) can be cast as a version of consequentialism. **Paul Hurley** in "The Consequentialist Argument Against . . . Consequentializing?" argues that this project faces a dilemma: either the consequentializer leaves out deontic evaluations central to other non-consequentialist views and thus fails to consequentialize the target theory, or in accommodating such evaluations fails to remain consequentialist. In developing his case, Hurley distinguishes two types of deontic evaluation of actions: *reason-independent* views that determine the deontic status of actions independently of whether the agent performs them for the right reasons, and *reason-dependent* views that take the agent's reasons for action as deontically relevant. In defending the first horn of the dilemma, Hurley argues that because consequentialist views standardly recognize only reason-independent conceptions of deontic status, they ignore views (like Kant's) that recognize reason-dependent conceptions. Hence, consequentializing does not work against Kant's theory and similarly for some versions of virtue ethics. However, if the consequentializer opts to recognize reason-dependent conceptions of deontic status, then the project of consequentializing fails to produce a version of consequentialism. Somewhat ironically, then, if Hurley is right then the consequentializing project constitutes an argument against consequentializing. Hurley concludes by reflecting on lessons that flow from his argument.

The concept of moral worth as typically employed in moral philosophy entails non-accidentally doing the morally right action. In the final chapter, "Moral Worth and Our Ultimate Moral Concerns," **Douglas W. Portmore** proposes a novel account of moral worth that relies on the idea of an ultimate moral concern, the "concern view" as he calls it. He begins the chapter by explaining that moral worth should be understood modally and so not just by reference to a person's motivating reason for performing a right action. Rather, the moral worth of an action performed on some circumstance depends on all and only those concerns that determine whether the agent acts rightly not only in the particular circumstance but in all other contextually relevant similar situations. He then argues by counterexample against the two leading theories of moral worth—the rightness view that can be traced back to Kant according to which an action has moral worth if and only if (and because) it is motivated by a non-instrumental concern for doing what is right, and the Hume-inspired right-making reasons view according to which an action has moral worth if and only if (and because) it is motivated by a non-instrumental desire to perform the action for the reasons that make an action right. (For the sake of easy comparison, we assume a Humean theory of motivation, not that versions of these theories must be so committed.) Portmore's diagnosis of the failure of these views is that they fail to accommodate that all and only those right actions "that issue from an appropriate set of concerns have moral worth." After arguing that a complete moral theory ought to specify such a concern, Portmore proceeds to draw a series of lessons from his concern account of moral worth, including that an action can have moral worth when it is not motivated by the thought that it is right, contra Kantian views, and contra Humean views that an action can have moral worth even though it is not performed for reasons that make the action right.

References

Broome, J. 1995. *Weighing Goods: Equality, Uncertainty and Time.* Oxford: Wiley-Blackwell.

Broome, J. 1999. *Ethics out of Economics.* Cambridge: Cambridge University Press.

Schroeder, M. 2019. "Persons as Things," in *Oxford Studies in Normative Ethics* 9. Oxford: Oxford University Press: 95–115.

1

Old-Fashioned Vices in Contemporary Crises; or, It Matters How You Value Yourself

Robin S. Dillon

During 2020–21, praise was frequently bestowed on people for virtues that enabled them to deal in ways often called "heroic" with the year's extraordinary health, economic, and political crises—praise, for example, for the courage and compassion of health care providers who put their own lives at risk to treat people infected with COVID-19; for the commitment of essential workers who kept important things operating, supply chains open, transportation running, and people fed; for the concern for justice expressed by tens of millions of people who participated in Black Lives Matter marches around the world; for the personal integrity of the U.S. state election officials who, despite disparagement and threats, resisted demands that they override the U.S. presidential election results.[1] But being somewhat misanthropic, I was more struck by the vices manifested in wave after wave of bad, even wicked, behavior, such as the greedy hoarding of toilet paper, hand sanitizer, and other items necessary for health and civilized existence; the egocentrically disrespectful refusals to wear face masks and socially distance to protect others from infection during the pandemic; the vicious anger, cruelty, and contempt of police and private individuals who attacked and threatened the lives of the Black Lives Matter protesters, not to mention the virulent rage and contempt of the thugs who, on January 6, 2021, invaded the U.S. Capitol, threatened the lives of government officials, and attacked democracy when Congress met to certify the electoral college results; the

---

[1] Tim Wu, "What Really Saved the Republic From Trump?" *The New York Times*, December 10, 2020 (https://www.nytimes.com/2020/12/10/opinion/trump-constitution-norms.html?searchResultPosition=1).

Robin S. Dillon, *Old-Fashioned Vices in Contemporary Crises; or, It Matters How You Value Yourself* In: *Oxford Studies in Normative Ethics, Volume 12*. Edited by: Mark Timmons, Oxford University Press. © Robin S. Dillon 2022.
DOI: 10.1093/oso/9780192868886.003.0002

political cowardice and sycophancy of Republicans who kowtowed before their idol's every outrage, coupled with their unbridled self-serving attempts to quash democracy by having their candidate for President declared the winner despite overwhelming evidence that he was the loser; and the non-stop, atrocious moral and legal crimes of the depraved, vice-filled asshole who occupied the White House until January, 2021.[2]

Despite their contemporary guise and the obvious differences among the vices I mentioned, and many more besides, my claim, which is an old-fashioned one, is that they share an important feature: they are all matters of distorted or perverse self-valuing. How old-fashioned is the claim? The idea that perverse self-valuing is at the heart of vices is a central tenet of the Deadly Sins tradition, which extends back to the fourth century CE.[3] And while Kant's account of vices is different in many ways from the Deadly Sins account, it shares the idea of deformed self-valuing as central to vices. What is more, both accounts hold that a particular form of perverse self-valuing underlies all vices, namely, what they both call "superbia": the sin of pride, the vice of arrogance.[4]

In this chapter I draw on the Deadly Sins and Kantian approaches to sketch an account of vice that centers the perverse self-valuing of arrogance, focusing on what might seem like the mildest of the vices in the above litany, namely, greed.

[2] I use the term "asshole" not only to express abundantly earned and justified contempt but also in the philosophical sense explored by Aaron James in *Assholes: A Theory* (Anchor Books, 2012) and *Assholes: A Theory of Donald Trump* (Doubleday, 2016). On justified contempt, see Macalester Bell, *Hard Feelings: The Moral Psychology of Contempt* (Oxford University Press, 2013).

[3] The most important traditional analyses of the Deadly Sins are those of the monks Evagrius Ponticus (4th century CE) and John Cassian (5th century CE); Pope Gregory I (7th century CE), whose analysis brought the scheme of deadly sins into official Christian doctrine (Gregory the Great, *Moralia in Job*, translated as *Morals on the Book of Job* by James Bliss and Charles Marriott (John Henry Parker and F. and J. Rivington, 1850)); and Thomas Aquinas (13th century), in *Summa Theologica* (Literally translated by Fathers of the English Dominican Province (Burns, Oates & Washbourne, 1921)) [cited below as *ST*]. Chaucer's "Parson's Tale," in the *Canterbury Tales* (14th century) is an extended sermon on the Seven Deadly Sins (*The Works of Geoffrey Chaucer*, ed. F. N. Robinson (Houghton Mifflin, 1933)). That the tradition is not dead is clear from Rebecca Konyndyk DeYoung, *Glittering Vices* (Brazos Press, 2009), and Gabriele Taylor, *Deadly Vices* (Clarendon Press, 2006).

[4] I've argued that arrogance is the heart of vices on the Kantian account in "Self-Respect, Arrogance, and Power," in *Respect for Persons,* ed. Richard Dean and Oliver Sensen (Oxford University Press, 2021); "Self-Respect and Humility in Kant and Hill," in *Reason, Value, and Respect: Kantian Themes from the Philosophy of Thomas E. Hill, Jr.,* ed. Mark Timmons and Robert Johnson (Oxford University Press, 2015); and "Kant on Arrogance and Self-Respect," in *Setting the Moral Compass: Essays by Women Philosophers,* ed. Cheshire Calhoun (Oxford University Press, 2003).

I. Preliminaries

A. Vice

The vices I am concerned with are morally bad traits of individuals' characters.[5] I take traits of character to be entrenched dispositional complexes of attention, thought, judgment, attitude, emotion, desire, and choice.[6] A full understanding of any character trait requires understanding these various dimensions. However, much can be gained by focusing on what I take to be the most significant aspect of the complex, namely, what I call one's "valuings": what one does and doesn't care about, how much and how much more or less, in what ways, and for what reasons,[7] all of which shape how one attends, perceives, interprets, judges, reacts, and is motivated to choose and act. The various character traits involve valuing many things; but I would argue both that self-valuing is central to each trait and that how one values oneself profoundly structures one's other valuings and thus the whole of one's character.[8] Distorted self-valuing, I hold, is an essential part of what makes a trait a vice, although the form of distortion could be different in different vices. If I am right, then the perverse self-valuing of vices is a grave matter indeed.

---

[5] One might wonder about the value of focusing on vices of individuals, rather than on institutional "villains," such as corporate consumer capitalism; entrenched systemic racism; media outlets that amplify, distort, and often excuse even blatant wrongdoing; the growing militarization of police power and institutionalized refusals to hold individual police officers accountable for mistreatment and deaths; or increasingly partisan politics. Attention to such things is indeed important; but so is attending to vices of individuals, as I argue in "Critical Character Theory: Toward a Feminist Theory of 'Vice'," in *Out from the Shadows*, ed. Sharon Crasnow and Anita Superson (Oxford University Press, 2012).

[6] I draw here on Hursthouse and Pettigrove, although I disagree with their identification of the most significant aspect of a character trait to be "the wholehearted acceptance of a certain range of considerations as reasons for acting" (Rosalind Hursthouse and Glen Pettigrove, "Virtue Ethics," *The Stanford Encyclopedia of Philosophy* (Winter 2018 Edition), Edward N. Zalta (ed.) (https://plato.stanford.edu/archives/win2018/entries/ethics-virtue/); sect.1.1).

[7] Some theorists treat valuing as regarding something as good in some way. But I use the term neutrally, in the sense of assigning or acknowledging some value, positive or negative.

[8] None of this is meant to deny that an individual might value herself in different ways at the same or different times, to the same or different degrees, at the same or different psychological levels; that there can be conflicts, even irresolvable conflicts, among one's valuings; or that someone might have vice V_1 and yet not have other vices. Humans, after all, are psychologically complex creatures.

B. Deadly Sins

While I am not prepared to argue for something like the inseparability of the vices, I do think that the Deadly Sins (or more accurately, deadly vices) approach is right in holding that individual vices are members of large and complexly interrelated families of vices.[9] In this tradition, the canonical Seven Deadly Sins of pride, envy, anger, greed, sloth, gluttony, and lust are regarded as the heads of various vice families. What makes these seven vices significant is not that they are or are supposed to be the most awful vices or most heinous wrongs a human can have or commit. Rather, their significance is reflected in the alternative designation of them as "cardinal" or "capital" vices: they are the most important (hence, cardinal) vices because they are regarded, individually and collectively, as the head(s) (L. *caput*) of all other vices and wrongdoings, or, as Aquinas says, the origin, "principle," and "leader."[10] The seven capital vices are, on this view, the most fundamental defects of character, the most basic ways humans go wrong.

Pride (*superbia*, arrogance) has long been regarded as having a unique status among the cardinal vices, as the root of the others. As Chaucer's Parson put it, although "everich of theise chief synnes hath his braunches and his twigges," pride is "the roote of these sevene synnes" and "the general roote of alle harmes."[11] Gregory calls pride the queen of the vices;[12] Aquinas identifies it as the beginning of all sin.[13]

The capital vices are the source of other vices and wrongdoings in the sense of being final causes: the ends they seek are "that for the sake of which" wrong actions are undertaken.[14] For example, people regularly lie, cheat, steal, and murder out of greed, and so greed (or, what one is greedy for) is a final cause of such actions. As Aquinas explains, "a capital vice is that which has an exceedingly desirable end, so that in his desire for it a man goes on to the commission of many sins, all of which are said to originate in that vice as

[9] In addition to the primary sources in note 3, see also Morton W. Bloomfield, *The Seven Deadly Sins: An Introduction to the History of a Religious Concept, with Special Reference to Medieval English Literature* (Michigan State College Press, 1952); Henry Fairlie, *The Seven Deadly Sins Today* (Notre Dame University Press, 1978); and Solomon Schimmel, *The Seven Deadly Sins: Jewish, Christian, and Classical Reflections on Human Psychology* (Oxford University Press, 1997).
[10] Aquinas, *ST* I-II, Q. 84, art. 3.
[11] Chaucer, "The Parson's Tale," Part III, lines 388–9.
[12] Gregory, *Morals* XXXI, ch. 87, p. 489.   [13] Aquinas, *ST* II-II, Q. 162, art. 7.
[14] *ST* I-II, Q. 84, arts. 3, 4.

their chief source."[15] The vices, that is, are a matter of perverse valuing: they involve valuing things in ways that lead one away from what is right and good.[16]

The perverse valuing is what makes the vices deadly. As Aristotle points out in the first lines of *Nicomachean Ethics* and Kant makes clear throughout his ethical works, it is constitutive of our being as human that we value. Each of the vices, then, involves a betrayal of our valuing capacity, damaging us at precisely the point in our nature that makes us human. Indeed, as Gabriele Taylor notes, they are "corruptive of the self," in the literal sense of corruption as destroying or dissolving the constitution that makes something the thing that it is.[17] Moreover, the deformed valuings the vices involve not only pervade and distort our whole character but also warp relationships with others and human community more generally. They alienate us from genuine value, including the real worth of ourselves and other people, lead us to violate even our most fundamental moral obligations, and put us deeply at odds with ourselves.

C. Self-Valuing

Jean Hampton argued in a series of essays that there are many different theories of human worth.[18] Such theories specify whether humans have intrinsic or only instrumental worth, whether human worth is inherent and natural or ascribed and socially constructed, whether people have equal or unequal worth, in virtue of what feature(s) someone has worth, and what the implications of having worth are for how to behave and to treat someone.[19] There thus are as many ways to value oneself as there are theories of the worth one might take oneself to have or lack. But I'll focus

---

[15] *ST* II-II, Q. 153, art. 4.

[16] *The American Heritage Dictionary*, s.v. "perverse": "1. Directed away from what is right or good; perverted."

[17] Gabriele Taylor, "Vices and the Self," in *Philosophy, Psychology and Psychiatry*, Royal Institute of Philosophy Supplement 37, ed. A. Phillips Griffiths (Cambridge University Press, 1994), 145–57; 145. See the *OED.*, s.v. "corruption."

[18] See especially Jean Hampton, "The Wisdom of the Egoist: The Moral and Political Implications of Valuing the Self," *Social Theory and Practice* 14 (1997): 21–51; and "Selflessness and Loss of Self," *Social Philosophy and Policy* 10 (1993): 135–65.

[19] Hampton, "Wisdom of the Egoist," pp. 27–9. Like Hampton, I take the Kantian conception to be the proper account of the moral worth of persons, and the proper account of morality to be one that has this conception of human worth at its heart (Hampton, "Selflessness," p. 147).

on three general forms of self-valuing: self-respect, self-esteem, and self-love. Each expresses a different theory of self-worth.

1. Self-respect

Self-respect is an appropriate and engaged appreciation of oneself as having significant worth. There are, as I have argued elsewhere, different kinds of self-respect. I focus here on two, which are varieties of what Stephen Darwall calls "recognition self-respect."[20]

What I call "moral interpersonal recognition self-respect" is the reflexive form of the respect that each person is morally owed and has a moral right to claim from every person.[21] That respect, moral interpersonal recognition respect, is practical acknowledgement from a moral point of view of persons as having the moral worth and status that all and only persons have just in virtue of being persons, and of the fundamentally moral equality of all persons. On Kant's account, it is as rationally autonomous moral agents that persons are ends in themselves with the intrinsic, absolute, incomparable, and equal worth and moral status that Kant calls "dignity,"[22] and are thereby both owed the respect of all other persons and morally required to respect themselves. The End in Itself formulation of the Categorical Imperative, on my reading, declares that our fundamental moral obligation is to accord all persons, oneself as well as others, moral interpersonal recognition respect.[23]

Moral interpersonal recognition respect for a person involves recognizing that this being is a person; appreciating that all persons have the distinctive moral worth and status of dignity; understanding that the equal dignity of

---

[20] Stephen Darwall, "Two Kinds of Respect," *Ethics* 88 (1977): 34–49; reprinted in Dillon (ed.), *Dignity, Character, and Self-Respect* (Routledge, 1995).

[21] I use the term "moral" to signal that there are also non-moral forms of recognition respect and self-respect, forms that regard people from nonmoral points of view as having significant but nonmoral worth. Kant gestures towards social interpersonal recognition respect when he refers to "the different forms of respect to be shown to others in accordance with differences in their qualities or contingent relations—differences of age, sex, birth, strength or weakness, or even rank and dignity, which depend in part on arbitrary relations" (Kant, *Metaphysics of Morals*, in Kant, *Practical Philosophy*, ed. and trans. Mary J. Gregor (Cambridge University Press, 1996) [cited below as *MM*]; *Akademie* pagination 6:468). See my discussions of kinds of respect in "Respect," *The Stanford Encyclopedia of Philosophy*, Edward N. Zalta (ed.) (https://plato.stanford.edu/archives/spr2018/entries/respect/); and of social forms of respect and self-respect in "Self-Respect, Arrogance, and Power."

[22] There is disagreement among scholars as to whether dignity is a worth or a status, and about which it is on Kant's account. I am agnostic on this question, so I treat it as both. See Oliver Sensen, *Kant on Human Dignity* (De Gruyter, 2011).

[23] I follow Wood in reading the End in Itself formulation as commanding respect for persons; Allen W. Wood, *Kant's Ethical Thought* (Cambridge University Press, 1999).

persons imposes moral constraints on one's attitudes, desires, and conduct, and submitting to those constraints as authoritative; and so, viewing, valuing, and treating this person only in ways that are appropriate to and morally due all persons.

Moral interpersonal recognition *self*-respect involves understanding and valuing oneself as a being with dignity and an equal person among equal persons with an absolute right to—and the moral authority to demand[24]—the same practical acknowledgement from all others of one's moral status and worth, as well as a categorical duty to acknowledge one's equal dignity in action and attitude; and living in light of this self-understanding and self-valuing. Individuals with moral interpersonal recognition self-respect regard certain forms of attitude and treatment from others as their due as a person and other forms as degrading and beneath the dignity of persons; and, other things being equal, they are not willing to be treated by others in ways that mark them as less than a person. But it is not only defensive self-valuing: because moral interpersonal recognition respect and self-respect are continuous, the latter is (as much as the former) a basis for moral solidarity and relationships of mutual support and care.[25]

Self-respect is not only about how one engages with others; it is also about how one engages with oneself as a person. What I call "moral agentic recognition self-respect" involves regarding the dignity one has as a person not only as calling for respect from others but also as demanding and constraining the exercise of one's own agency, as giving rise to a responsibility to shape oneself and direct one's living so that they are congruent with and honor one's dignity as a person. An individual with moral agentic recognition self-respect values herself appropriately by committing to live in accord with norms that configure the kind of life she rightly regards as befitting her as a moral agent, and by, among other things, eschewing forms of acting, thinking, feeling, and desiring that are degrading, shameful, or beneath her as a person.

2. Self-Esteem

I use the term "self-esteem" as it is used by contemporary American psychologists and social psychologists, to refer to a positive, self-approving

---

[24] Stephen Darwall, *The Second-Personal Standpoint: Morality, Respect, and Accountability* (Harvard University Press, 2006).

[25] See my "Toward a Feminist Conception of Self-Respect," *Hypatia* 7 (Winter 1992): 52–69.

attitude—thinking well of or feeling good about oneself.[26] The worth that individuals with self-esteem ascribe to themselves can be intrinsic or instrumental, natural or socially relative. The assessment is socially comparative in that it involves one's view of one's acceptability to others and self-evaluation by social standards, and it is usually caused and sustained by acceptance and praise from others. Self-esteem also tends to be competitive self-valuing. People typically want to be more popular, more accomplished, more powerful, or have more or better than others; so self-esteem ascribes unequal worth to people and is higher the more one believes one outdoes others.

The possible grounds of self-esteem are very wide: they include anything about the self that the individual judges or believes that others judge to be valuable in some way. Importantly, there are no moral constraints on the grounds; one can derive high self-esteem from successful lying, cheating, and manipulation of others, or from breaking into the Capitol or looting Nancy Pelosi's office. Moreover, there is no intrinsic and necessary connection to reality; indeed, robust empirical evidence indicates that most adults have unrealistically high self-esteem.[27]

This is in large part due to what social scientists call "the self-esteem motive," which comprises an array of self-serving strategies (such as selective attention, rationalization, reinterpretation, and denial) that protect or boost positive self-esteem through a distorted view of the self.[28] In addition to enabling people to maintain or increase self-esteem in defiance of the evidence, it also makes people quite competitive for relative social worth and moves them to ascribe greater worth to themselves and their own qualities than to other people and their (similar) qualities. Most researchers accept it as an axiom that the desire to maintain and enhance self-esteem is an enormously powerful, ubiquitous, even universal motivation (albeit one whose operation is shaped by social and cultural context).

---

[26] See Christopher J. Mruk, *Self-Esteem Research, Theory, and Practice: Toward a Positive Psychology of Self-Esteem*, 3rd ed. (Springer, 2006); and essays in *Self-Esteem Issues and Answers: A Sourcebook of Current Perspectives*, ed. Michael H. Kernis (Psychology Press, 2006). Although self-esteem is commonly viewed as a positive attitude, as I treat it here, social scientists regard it as a spectrum attitude that ranges from very positive to very negative, as reflected in the familiar terms "high self-esteem" and "low self-esteem."

[27] Shelley E. Taylor and Jonathan Brown, "Illusion and Well-Being: A Social Psychological Perspective on Mental Health," *Psychological Bulletin* 103 (1988): 45–63.

[28] Viktor Gecas, "The Self as Social Force," in *Extending Self-Esteem Theory*, ed. Timothy Owens, Sheldon Stryker, and Norman Goodman (Cambridge University Press, 2001), 85–100; 87. See also Howard B. Kaplan, "Prevalence of the Self-Esteem Motive," in *Self-Attitudes and Deviant Behavior*, ed. Howard B. Kaplan (Goodyear, 1975); Timothy J. Owens and Alyson R. McDavitt, "The Self-Esteem Motive: Positive and Negative Consequences for Self and Society," in Kernis, *Self-Esteem*, pp. 398–406.

3. Self-love

While self-respect and self-esteem have to do with one's worth, self-love, as I'll use the term, has to do with one's well-being. I follow Kant generally, though not in detail: as the essence of love of others is practical benevolence, which involves "making the well-being and happiness of others my end,"[29] so *self*-love is "predominant *benevolence* toward oneself."[30]

Considered in itself, there is nothing disordered or perverse in valuing one's own well-being. Indeed, concern for one's well-being serves an end that is, other things being equal, good, death and misery being among the bad results of deficient concern. But self-love can be disordered in, among other ways, how the concern for one's well-being possesses one and affects one's attention, judgments, attitudes, desires, and choices; in the prioritizing of one's own well-being over other things of value, including (but not only) the well-being of other people; and in what one is willing to do to secure or advance one's well-being.

Forms of disordered self-love include the kind that Kant is most concerned with, namely, desires or willings that are not constrained by one's acknowledgment of the authority of the moral law, so that in seeking to advance one's self-interest one is prepared to do what is wrong or tolerate what is bad. Prioritizing reasons of self-love over moral reasons is the essence of the form of evil that Kant calls depravity, in which "the mind's attitude is corrupted at its root."[31] Other forms of perverse self-love, such as selfishness, self-centeredness, egoism, and self-indulgence, might seem less than evil, but can also be sources of vices and wrongdoing.

II. Greed

Greed is typically understood as an excessive desire for wealth or things that money can buy.[32] *The American Heritage Dictionary* defines it as "an

[29] *MM*, 6:452.
[30] Kant, *Critique of Practical Reason,* in Kant, *Practical Philosophy,* 5:73.
[31] Kant, *Religion Within the Boundaries of Mere Reason,* trans. George di Giovanni, in *Religion and Rational Theology,* ed. Allen W. Wood and George di Giovanni (Cambridge University Press, 1996), 6:30.
[32] There are, interestingly, rather few contemporary philosophical examinations of greed; most focus on business ethics. General accounts include DeYoung's *Glittering Vices*; Andrew Pinsent, "Avarice and Liberality," in *Virtues and Their Vices,* ed. Kevin Timpe and Craig A. Boyd (Oxford University Press, 2014), pp. 157–75; and Jason Kawal, "Rethinking Greed," in *Ethical Adaptation to Climate Change,* ed. Allen Thompson and Jeremy Bendik-Keymer (MIT Press, 2012), pp. 223–39.

excessive desire to acquire or possess more than one needs or deserves, especially with respect to material wealth." The *OED* defines it as "inordinate or insatiate longing, especially for wealth," offering craving, intensity of desire, and rapacity as characteristics. But since one can be greedy for things other than wealth, like attention, power, or status, and since we call greedy those who take more than their share of the pizza or eat like pigs (both shoveling and eating indiscriminately), we could say that on the typical view greed involves excessively desiring or going for things in excess: the desire or going for is inordinate, and what one desires or goes for is much more than what one needs or deserves.

There are different kinds of greed. Kant's chief concern is with miserliness: willing to accumulate goods with no intention of making use of them, leaving one's "true needs unsatisfied."[33] Another is what Aristotle seems chiefly concerned with, stinginess: being disposed to give too little to others (or giving, but being pained by it), typically because one "take[s] wealth more seriously than is right."[34] *Pleonexia*, as Irwin glosses it, is another: the "desire to have more than I am entitled to, so as to get the better of someone else."[35] I want to consider greed in the context of amassing goods made scarce as a result of the pandemic.

In March and April of 2020, as COVID-19 spread across the U.S., lots of people engaged in what was called "panic-buying" of toilet paper. The first thing I encountered on my first visit to a supermarket after weeks of sheltering-in-place was a man pushing a cart piled so high with packages of toilet paper that it was almost impossible to see a human behind the cart. When I got into the store, I found the toilet paper shelves completely empty. Newspapers and websites were filled with reports of people across the country grabbing as much toilet paper as they could and of people exchanging harsh words, shoves, and even blows over others' grabbing (being thwarted, one supposes, in their own attempts to grab). Now, the individual I saw might have been buying for the local homeless shelter, or he could have been the designated buyer for his extended family or his neighbors;[36] but

---

[33] *MM*, 6:432.

[34] Aristotle, *Nicomachean Ethics*, trans. Terence Irwin, 2nd ed. (Hackett Publishing Company, 1999), 1119b30.

[35] *NE*, Glossary, "overreaching, *pleonexia*," pp. 340–1.

[36] Could greed be selfless? On some accounts, greed is necessarily and exclusively self-centered. However, just as pride, a form of self-valuing, can extend to being proud of one's children or of some group with which one identifies, so I think the self-love that greed involves can extend to others with whom one identifies, in the sense of seeing one's own well-being as bound up with theirs: one might want to amass lots for the sake of loved ones. I thank Joseph Bowen for pressing this question.

those possibilities likely did not apply to most people who were snatching up as much toilet paper as they could.

Now, there is clearly something wrong about coming to blows over toilet paper. But in normal circumstances it seems that there isn't anything morally objectionable about buying in bulk. Indeed, not only are some U.S. stores designed specifically for bulk buying, but regular stores frequently encourage buying in quantity (such as the "10 for $10" sales that my local supermarket regularly runs). Nor is there an obvious moral problem with buying what one wants and not leaving enough for others. In many cases, shoppers know that there are limited quantities (so we are advised to shop early for best selection); and when the item we want is out of stock, while we might be annoyed, I think we tend not to criticize the folks who got there before we did. In unusual cases, such as when a hurricane or major snowstorm is approaching, people tend to stock up on items they think they'll need; but we typically don't think of people who buy extra batteries or toilet paper in anticipation of being stuck at home during a nor'easter as selfish or greedy. So, is there anything morally objectionable about buying large quantities of goods during a pandemic? And is it even greed?

One might be inclined to think that the cause of this buying in excess was not greed, but something else. The term "panic-buying" suggests that the cause was fear—fear that one might not have enough toilet paper if one were confined to the house for weeks. News reports about the toilet paper shortage offered other explanations, including guilt avoidance (people don't want to be the one who doesn't have what they need), social cues (if others are buying a lot, I should do likewise), and anxiety management (I can't control COVID-19, but I can control what I buy). One study claimed that certain personality factors explained who was panic-buying toilet paper: "people who feel more threatened by COVID-19 and whose personality is characterized by a particularly high degree of emotionality and conscientiousness were more likely to stock up on toilet paper than people who do not have these characteristics."[37] Nevertheless, I think that greed was still involved.[38]

---

[37] "The psychology behind toilet paper hoarding," *MyScience.org*, June 15, 2020 (https://www.myscience.org/en/news/2020/the_psychology_behind_toilet_paper_hoarding-2020-uni-muenster).

[38] Of course, not all grabbing of so much manifests greed, for someone might be acting out of character. One way to tell is by the individual's reflective responses, e.g., whether they feel ashamed or guilty afterwards, or resolve to exercise more self-control next time.

It is not, however, useful to think about what's going on with toilet paper hoarding as the typical view of greed would have it, as either desiring to possess toilet paper in excess of what one needs or as an insatiable craving for toilet paper. When people had no way of knowing how long they would have to forgo trips to the store, or how long it would be until the shelves would be restocked, it might have seemed reasonable, or at least not unreasonable, to buy lots of toilet paper to make sure that one had enough for the duration. And I'm not sure we can even understand an intense or insatiable desire for toilet paper as anything other than a sign of some psychological problem.[39]

I think the Deadly Sins approach has a better explanation of greed: it is, on one level, a matter of the disordered valuing of things that one believes or hopes would contribute to one's well-being and of how one's valuing of such things affects and is affected by one's other valuings, judgments, attitudes, emotions, and motivations; more deeply, it is a matter of disordered self-love. Self-love can be disordered, in addition to the ways mentioned above, in one's emotional responses to opportunities to give or get; in one's judgments about how much is enough, one's view of need, and one's competitiveness; in one's reasons for wanting more; in mindless habits of acquisition and possession; in the ways self-esteem and even self-definition hinge on what one can call "mine." The Deadly Sins analysis sees the various vices as operating not only as direct motivations, as, for example, a felt insatiable desire, but also by giving rise to or reinforcing other motives, or by shutting down or overriding motivations or reasons that might prevent one from acting wrongly,[40] or by distorting judgment and valuing. Fear of running out of toilet paper might have been what some people hoped to allay; conscientiousness might have been what led others to find it reasonable to stockpile toilet paper; but greed was likely the deep motivation for both the excessive buying and the willingness to come to blows over toilet paper.

Why do I say "likely the motivation?" Because, as I hinted above, greed is normalized, sustained, and reinforced by the sociocultural arrangements and operations of consumer capitalism.[41] Consumer capitalism requires that people buy more and more of whatever is offered, and so it trains us

---

[39] I thus agree with Kawall that greed is not essentially a matter of desire intensity or of the objects or amounts one desires.

[40] DeYoung describes *pleonexia* as "the absence of a certain restraint on the desire for gain" (*Glittering Vices*, p. 192).

[41] On the sociocultural background for greed, see, e.g., Amelie Oksenberg Rorty, "Political Sources of Greed and Anger," *Midwest Studies in Philosophy* 22 (1998): 21–33; and Dorothy

in a multitude of ways to want more and more, to think that we really *need* the things we want so that we are highly motivated to buy more and more, to buy even in excess of our perceived needs, to buy because "I deserve it," or "it's a bargain," or as a form of "therapy," and to think of ourselves as entitled to buy whatever we can afford to buy. But while we are socially encouraged and trained to be greedy, we are, nevertheless, morally responsible for how we handle social encouragement and training; and so, people who desire and act as they are socially encouraged to do are still criticizable as greedy.

The heart of the disordered valuing that greed involves is perverse self-love: a morally objectionable concern for one's own well-being. But what makes the concern morally bad? There are at least two objectionable features of greedy hoarding.

The first feature is the way greed involves prioritizing one's well-being over that of others. I assume that ordinarily, in most parts of the United States, people are used to buying whatever they want or think they need,[42] without giving a thought to what others might want or need, because "normal" shopping does not require us to think about others—in our usual experience, there is enough left for the next person; or if there's not, well, there's not an unlimited supply of anything one could buy, and somebody has to get the last one. Yet it takes little thought to figure out, when I am shopping as the nor'easter is bearing down, that if I take all the toilet paper or soup on the shelf now, it might not be replaced right away, maybe not for many days, and then others won't be able to buy what they need. But, for many people, the habits of buying that consumer capitalism trains into us, and the typical lack of competition among the people shopping in "normal" conditions, impede this easy thinking, reinforcing the general tendency to think only about oneself and to buy as much as one wants, regardless of others. One problem with pandemic hoarding is that taking so much more than one's fair share without leaving "enough and as

L. Sayers, "The Other Six Deadly Sins," in Sayers, *Creed or Chaos?* (Methuen, 1947). Alan Johnson's Monopoly metaphor is useful for thinking about the normalization of greed. The game Monopoly encourages and rewards greed: the point is to amass money and property at the expense of others, and whoever outdoes all the other players wins. The people whose greediness in real life we criticize are like really good, i.e., really ruthless, Monopoly players, who are doing exactly what the game requires them to do, just rather more extremely. Taking such folks to be paradigms of greediness helps us not to see the ways in which we, too, are greedy, albeit (perhaps) to a lesser (normalized) degree. Alan G. Johnson, *Privilege, Power, and Difference* 2nd ed. (McGraw Hill, 2006), pp. 82–83, 86.

[42] I am conscious of the class privilege reflected here.

good" for others, especially of things that everyone needs and that are in short supply and that one knows are in short supply, means that others must do without things that are needed for health and civilized existence. So, we can understand greedy taking as a form of injustice, as Aristotle does.[43] But the tendency to injustice arises from prioritizing one's own well-being to such an extent that the well-being of others does not register as a significant enough reason (if a reason at all) to restrain oneself from grabbing as much as one can, as if we were back in the state of nature where anything goes. The first objectionable feature of greed, then, is that it involves acting as if other people are morally irrelevant—and this is true even in "normal" conditions.

Another case of greedy hoarding highlights a second objectionable feature. *The New York Times* reported that in March, 2020, two brothers, Matt and Noah Colvin, had driven over a thousand miles across Tennessee and Kentucky, buying all the remaining bottles of hand sanitizer and packages of antibacterial wipes, products regarded as vital to fighting the spread of COVID-19.[44] The Colvins ended up with 17,700 bottles of hand sanitizer, which they planned to sell on Amazon for prices ranging from $8 to $70 a bottle, thus potentially earning close to a million dollars.[45] Nor was this the first time: a month earlier, the article reported, Matt Colvin sold $700 worth of face masks for $80,000–$100,000;[46] and "the success stoked his appetite. When he saw the panicked public starting to pounce on sanitizer and wipes, he and his brother set out to stock up."[47]

---

[43] As Aristotle says, "the unjust person is an overreacher [i.e., someone possessed by *pleonexia*]" (*NE*, 1129b2). In the glossary of his translation of *NE*, Irwin notes that "*pleonexia* always involves injustice" (p. 340). DeYoung agrees: "The greedy are excessive in acquiring and keeping possessions even to the point of depriving others of what they deserve or need ... Justice ... is the virtue it attacks" (*Glittering Vices*, p. 107).

[44] Jack Nicas, "He Has 17,700 Bottles of Hand Sanitizer and Nowhere to Sell Them," *The New York Times*, March 14, 2020 (https://www.nytimes.com/2020/03/14/technology/corona virus-purell-wipes-amazon-sellers.html).

[45] According to Nicas, "Mr. Colvin is one of probably thousands of sellers who have amassed stockpiles of hand sanitizer and crucial respirator masks that many hospitals are now rationing ... Amazon said it had recently removed hundreds of thousands of listings and suspended thousands of sellers' accounts for price gouging related to the coronavirus."

[46] This is my estimation of the gross receipts, not net profit.

[47] Lest I be misunderstood: My purpose in discussing this case is not to assess the Colvins' characters, for I know nothing about them other than what I read in two news reports. Moreover, it is important to be cautious and generous when addressing the characters of other individuals (and even of oneself). For there can be, maybe even typically are, multiple reasons and motivations at work in how an individual acts in a certain context; those reasons and motivations can be conflicting or unconscious and even resistant to self-discovery; and particular situations can themselves be complex. What is more, whether a particular behavior manifests a particular vice can depend on other values, aims, and traits that an individual has. It thus behooves one to acknowledge that assessments of character and actions are always fallible.

The problem highlighted by actions like the Colvins' is not the inordinate desiring or rapacious taking in excess of their own needs. Nor is the problem that they or other folks who were amassing stockpiles of health-protecting items and engaging in price-gouging thought that other people's well-being was irrelevant; indeed, other people's desperation was essential to their plans to make a huge profit. But such folks could not have been thinking of prospective buyers as *people*, as fellow human beings whose unmeetable needs to protect themselves from COVID-19 should have evoked compassion.[48] It seems more likely that they saw potential buyers only as "marks" whose willingness to pay exorbitant prices could be very profitably exploited. The second objectionable feature of greed is that the prioritizing of their own well-being can involve commodifying the well-being of other people, who are treated merely as means to the exploiters' end of making a million bucks.

Both treating persons merely as means and treating them as morally irrelevant, which the disordered self-love of greed engenders, are disrespectful of other persons. They are also self-disrespectful.

To return briefly to the other forms of greed I mentioned above: miserly greediness, stinginess, and *pleonexia* are also matters of disordered self-love that also could result in depriving and disrespecting others. And insofar as the motivation to get more and more is to "get the better of someone else," greed is also a matter of disordered self-esteem.

III. Disrespect

Disrespect of others involves the failure or refusal to regard and treat them with moral interpersonal recognition respect; disrespect of oneself involves, in this context, the failure of either moral interpersonal recognition self-respect or moral agentic recognition self-respect. Consider first disrespect of others.

The theory of worth underlying moral interpersonal recognition respect holds that all persons have equal worth and standing in the moral community and are equally worthy of respect. Moral interpersonal recognition respect for persons affirms the *categorical* constraints on attitudes and conduct that dignity places on us, acknowledging that we are morally

---

[48] In the Deadly Sins tradition, greed is said to give rise to what Lyman calls "hardness of heart against compassion" (Stanford M. Lyman, *The Seven Deadly Sins: Society and Evil* (St. Martin's Press, 1978), p. 235.

obligated *always* to respect persons and *never* to treat them merely as a means or as if they had no moral importance whatsoever. More specifically, as Kant explains, the duty of respect for others is "to be understood as the *maxim* of limiting our self-esteem by the dignity of humanity in another person."[49] The duty of respect is a negative one "of not exalting oneself above others," which is "contained in the maxim not to degrade another to a mere means to my end" and which requires that "I keep myself within my own bounds so as not to detract anything from the worth that the other, as a human being, is authorized to put upon himself."[50]

There are, unfortunately, very many ways of disrespecting persons. Of particular relevance to greed are regarding or treating persons as if they were not one's moral equals but occupied a lower level of a hierarchically organized moral community; as if they were not ends in themselves with dignity but merely means with only instrumental value insofar as they serve one's interests, including one's well-being or self-esteem; as if they were not sources of moral constraints on one's actions but as if they and their well-being were morally irrelevant and so things that one does not have to take into consideration in deciding how to act.

Moral interpersonal recognition *self-respect* rests on an understanding of the moral community as wholly composed of equal persons; so, the self-respecting person values herself as a being with equal dignity who is inescapably related as an equal person to all other persons. To engage with others as if the moral community were hierarchical and as if one were morally more important, intrinsically more valuable than others (or less than others, for that matter) is to fail to acknowledge one's basic moral relatedness with others, to deny that one is a being whose most fundamental and important worth and status is *equal* dignity, and to deny that one is an equal person among equal persons. It is thus to lack moral interpersonal recognition self-respect.[51]

What Kant calls the "slavish subjection of oneself to the goods that contribute to happiness"[52] by failing to be the master of one's desires and emotions is a failure of moral agentic recognition self-respect. As Kant explains, when someone acts on maxims that derive their motivating power from inclination rather than reason, he "deprive[s] himself of the

---

[49] *MM*, 6:449.   [50] *MM*, 6:450.
[51] I explain this more fully in "Kant on Arrogance and Self-Respect," "Self-Respect and Humility in Kant and Hill," and "Arrogance, Self-Respect and Personhood," *Journal of Consciousness Studies* 14 (2007): 101–26.
[52] *MM*, 6:434.

prerogative of a moral being, that of acting in accordance with principles, that is, inner freedom … the innate dignity of a human being," in effect "mak[ing] it one's basic principle to have no basic principle" and so making oneself "a plaything of the mere inclinations, and hence a thing," and, as it were, "throw[ing] oneself away."[53] To let oneself be controlled by one's inclinations—whether by one's desires for more and more stuff or ever greater self-esteem, or by one's fear, guilt, anxiety, or whatever—is to make oneself a plaything of them, and so is to fail to respect oneself as a rational moral agent.

These ways of disrespecting others and oneself are expressions of the distorted self-valuing that is the heart of arrogance.

IV. Arrogance

I have argued elsewhere that there are at least two kinds of arrogance, which I call "status arrogance" and "unwarranted claims arrogance." Although they comprise different dispositional complexes of cognition, attitude, and desire, as vices their common core is an orientation to self-worth that is fundamentally flawed.[54]

Consider first status arrogance. I expect that most of us think of arrogance as essentially interpersonal, a matter of how someone treats other people.[55] The arrogant individual is generally thought to be someone who thinks he is better than other people, who looks down on others and treats them contemptuously, peremptorily, or without consideration, making it clear that he views them as less important, less valuable than his very important, very valuable self. This view of arrogance is expressed in the definition given in *The American Heritage Dictionary*: "a sense of overbearing self-worth or self-importance, marked by or arising from an assumption of one's superiority toward others." Status arrogance can begin in the belief that one's abilities or accomplishments are greater than those of others or that one has a higher social status. But its hallmark is having a settled conception of

---

[53]   *MM*, 6:420.

[54]   I argue in "Self-Respect, Arrogance, and Power" that unwarranted claims arrogance is not always a vice but can sometimes express morally proper recognition self-respect, in which case the self-valuing is morally justified, not distorted.

[55]   See, e.g., Tiberius and Walker, who argue that arrogance is essentially "an interpersonal matter. It consists in a particular way of regarding and engaging in relations with others." Valerie Tiberius and John D. Walker, "Arrogance," *American Philosophical Quarterly* 35 (1998): 379–90; 381.

oneself as superior in the sense of having both a greater worth or importance and a higher normative status than others. The status-arrogant individual takes himself to be entitled to Good Things, such as rights, privileges, respect, and precedence, to which others have no legitimate or equally important claim (although he may generously grant these undeserved things to them). And he takes himself to be entitled to treat others as inferiors, to make demands on them and expect their deference, to insist that his needs and wants take priority over theirs, or to dismiss or ignore them as not worth his consideration.[56] Moreover, the status-arrogant person not only values himself highly but also highly values having superior worth and status. And both the self-valuing and the valuing of self-valuing are disordered: inordinate and morally unjustified.

Kant develops a view of status arrogance as involving an unjustified belief in one's superior worth that "demands from others a respect it denies them."[57] As he explains, "arrogance (*superbia* and, as the word expresses it, the inclination to be always on top) is a kind of ambition (*ambitio*) in which we demand that others think little of themselves in comparison with us ... Arrogance is, as it were, a solicitation on the part of one seeking honor for followers, whom he thinks he is entitled to treat with contempt."[58] This presumptuous "lack of modesty in one's claims to be respected by others"[59] is "a vice opposed to the respect that every human being can lawfully claim."[60]

Status arrogance violates the duty of moral interpersonal recognition respect, for it involves treating others as beings of a lesser kind with little worth and no claim to respect as equal persons. Those afflicted with status arrogance also fail to have moral interpersonal recognition self-respect, for they exalt themselves over others, embracing a hierarchical theory of the worth of persons and locating themselves on the upper tier, rather than valuing themselves as equal persons among persons.

Consider now unwarranted claims arrogance. While the *American Heritage* definition stresses how one regards and interacts with others in

---

[56] While the conception of one's more worthy self can be idiosyncratic, it can also be part of one's identity as a member of a certain group, class, or people—'our kind' as opposed to 'those others:' the arrogant heart of racism, sexism, religious animosity, and other forms of bigotry and oppression. Moreover, the arrogant needn't think of themselves as superior to everyone else, and arrogance with respect to some people is compatible with acknowledged inferiority to other people, indeed can express, compensate for, or protect against a sense of inferiority. But status arrogance can also be just what it appears to be: the manifestation of absolute confidence in what one takes to be one's objectively greater worth and higher normative status.

[57] *MM*, 6: 465.   [58] *MM*, 6: 465.   [59] *MM*, 6: 462.   [60] *MM*, 6: 465.

light of how one values oneself, the OED definition reminds us that the adjective "arrogant" derives from the verb "to arrogate:" "to assume as a right that to which one is not entitled; to lay claim to and appropriate (a privilege, advantage, etc.,) without just reason or through self-conceit, insolence, or haughtiness (from L., to ask or claim for oneself)." Arrogance, then, is "the taking of too much upon oneself as one's right; the assertion of unwarrantable claims in respect of one's own importance; undue assumption of dignity, authority, or knowledge."[61]

Unwarranted claims arrogance involves a disposition to overreach: aspiring to what is beyond what one can legitimately claim. Many things towards which the disposition to arrogate is directed are what a person of higher rank or greater importance could rightly claim; appropriating them is a means of elevating self and self-worth. So, unwarranted claims arrogance can be powered by the self-esteem motive. This form of arrogance is opposed to moral agentic recognition self-respect, the proper acknowledgment and valuing of oneself as an agent that involves, among other things, taking seriously the moral responsibilities of agency. For Kant, our most vital responsibility is to realize our capacity for autonomous agency by choosing to act on rational motives, including acknowledging the absolute authority of the moral law, which is to say, the dictates of one's reason unimpeded by the importuning of inclination. The prioritization of self-love or the desire for self-esteem over reason's moral law involves the arrogation by one's inclinations of supreme moral authority and control of the will. When one subordinates rational autonomy to inclination, one makes oneself "a plaything of the mere inclinations," debasing one's dignity as a rational being and failing to respect oneself as a rational moral agent.

V. Arrogance, Disrespect, and Greed

Greed involves disrespect of other persons and oneself and both kinds of arrogance. The perverse self-love at the core of greed involves not merely wanting and taking lots for oneself, maybe more than one reasonably needs. It involves regarding one's needs and wants as unconditionally more

[61] The accounts in Frye and in Roberts and Wood are accounts of unwarranted claims arrogance. Marilyn Frye, "The Arrogant Eye," in *The Politics of Reality: Essays in Feminist Theory* (Crossing Press, 1983); Robert C. Roberts and W. Jay Wood, "Humility and Epistemic Goods," in *Intellectual Virtue: Perspectives from Ethics and Epistemology*, ed. Linda Zagzebski and Michael DePaul (Oxford University Press, 2003), pp. 257–79.

important than the needs and wants of others, and, more deeply, taking oneself to matter more, in an absolute sense, than other people.[62] The greedy not only give no thought to the needs of others; they act as if others are not worthy of any such consideration. The unwillingness of the greedy to restrict their self-love out of consideration for the self-love of others[63] shows that they exalt themselves above other persons and act out of the false hierarchical view of the worth of persons that status arrogance involves. Moreover, the fact that many greedy people were willing to come to blows over who gets that package of toilet paper indicates that, like other status-arrogant folks, they demanded recognition respect, in this case for their right to buy whatever and however much they wanted without interference from others, while denying other people the same respect.

The intention to exploit others by cornering the market on things that people genuinely need, like hand sanitizer or face masks, and then overcharging those desperate enough to pay any price, likewise shows status arrogance: the exploiters exalt themselves over other people, regarding them as things that can be used merely as tools for advancing the exploiters' well-being, while regarding only themselves as beings whose desires should be respected and fulfilled. The greedy thus treat others with contempt.

Greed also involves unwarranted claims arrogance. The greedy arrogate precedence for their well-being over the well-being of others, implicitly claiming a right to take as much as they want or to corner the market in needed items and then charge whatever they can get. But their claims are unwarranted. They act on maxims that derive their motivating power from unconstrained self-love rather than reason. And in slavishly subjecting themselves to their desire for toilet paper or profit, or fear or anxiety, they allow their inclinations to arrogate priority over their reason, making themselves playthings of their inclinations.

The self-esteem motive could be part of what drives the greedy. The hand sanitizer brothers might have thought that they were so much more clever than ordinary folks in figuring out early on that hand sanitizer was going to be in big demand and in figuring out how to amass lots of it. Thinking of oneself as smarter than others and being able to outdo others in buying and

---

[62] More generally, greed involves taking one's well-being to matter more than anything else of genuine value: more than other forms of life, the environment or planet, good character, justice, truth, and so on.

[63] MM, 6:462.

selling are common sources of self-esteem and can reinforce status arrogance and express unwarranted claims arrogance.

It should be clear, then, why greed is a deadly vice. It involves a self-destructive corruption of one's valuing capacity, inasmuch as it prioritizes distorted self-love and self-esteem over morally justified self-respect. Moreover, the more the greedy person lets themselves be controlled by egoistic self-love, the self-esteem motive, or passions like fear, the less they value themselves properly as rational beings capable of governing themselves through rationally justified principles (and not just maxims like "I'll take everything I can and to hell with everyone else"), which puts themselves at odds with themselves. The greedy are cut off from the real value of other people, and they violate the most fundamental and categorical moral duties of respect and self-respect.[64] The arrogant self-valuing that greed involves makes it not just one among many annoying traits but a deadly vice. The contemporary crisis that brought it to the fore was not only one involving a pandemic-induced shortage of goods; it was a crisis of distorted self-valuing and self-destruction.[65]

---

[64] They also violate the fundamental moral duty of love of others, which I treat as a species of respect for persons.

[65] I am grateful to participants at the Arizona Workshop in Normative Ethics for helpful and generous comments and discussion.

2

Robust Rights and Harmless Wronging

Joseph Bowen

1. Introduction

Rights are important. Part of why rights have this importance is that they protect us from harm—they protect our interests. This chapter examines a range of cases in which it appears one's rights against harm have been violated by another's behaviour, even though this behaviour has done no harm. Call these cases of "harmless wronging".

Plane Crash (Preempted Harm). Passenger is about to board a plane. Attendant takes a disliking to Passenger, so denies her admittance onto the plane. On departure, the plane crashes and everybody on board dies.

Roulette (Pure Risk Imposition). Target is asleep. Her housemate, Shooter, comes into her room and decides to play Russian roulette with her. Luckily, no bullet is fired. Shooter, content with having had a round of roulette, will never play roulette again.

Intuitively, Passenger and Target have their rights violated.[1] However, Passenger is better off in the world in which her rights are violated than that in which they are not. Target's life is as it would have been had Shooter not made Target the subject of her risky behaviour.[2] Given the standard

---

[1] An anonymous reader for this volume would like to hear more about why it is intuitive that there are right violations in *Plane Crash* and *Roulette*. Here is another way at getting at the intuition. Many hold that one is wronged by some action only if one holds a right against that action. I imagine Passenger would not think, merely, that Attendant acted wrongly, nor do I believe Target would think Shooter merely acted wrongly (were she to find out what happened). Rather, they would feel *wronged*. This suggests, assuming the stipulated connection between rights and wrongs, Passenger and Target have their rights violated.

[2] Some think that risk of harm is *itself* harmful, e.g. Finkelstein (2003). Elsewhere, I argue this is mistaken (Bowen Forthcoming).

Joseph Bowen, *Robust Rights and Harmless Wronging* In: *Oxford Studies in Normative Ethics, Volume 12.*
Edited by: Mark Timmons, Oxford University Press. © Joseph Bowen 2022.
DOI: 10.1093/oso/9780192868886.003.0003

Counterfactual Account of Harm, on which *Y* harms *X* iff *Y* makes *X* worse off than *X* would have been had *Y* not acted as she did, Passenger and Target are not harmed by the violation of their rights.

Cases of harmless wronging raise a serious problem for most theories of rights (namely, those theories that have wellbeing play some grounding role in rights). For example, most theories say that, other things being equal, the stringency of a right corresponds to the harm that would befall its holder were that right to be violated (Thomson 1990, 149–75; Kamm 2007, 249–75). Cases of harmless wronging wreak havoc with the intuitively plausible stringencies of rights. What is the stringency of Passenger's and Target's rights against Attendant's and Shooter's behaviour?

The problem caused by cases of harmless wronging is most pronounced given the dominant theory of the *nature* of rights, the Interest Theory of Rights. The problem is most pronounced for the Interest Theory because it is the theory that most depends on the intuitive connection between rights and harm that cases of harmless wronging call into question.

Interest Theory (Canonical). X has a right against *Y* that *Y* Φ, only if (and *because*) *X*'s wellbeing (her interests) is of sufficient weight to place *Y* under a duty to Φ.[3]

The Interest Theory has a lot going for it. As well as having an intuitively plausible account of the grounds of rights, it explains some fundamental structural features of rights and their correlative directed duties. First, it gives a good account of why *Y owes* her duty to *X*—the duty exists because of *X*. Second, it explains why, through infringing her duty, *Y* does not merely act wrongly but *wrongs X*—*Y* has failed to respond to morally salient duty-grounding features about *X*.

The Interest Theory runs into trouble with cases of harmless wronging. Neither Passenger's nor Target's wellbeing is protected by their putative rights that are violated. Because of this, it is hard to see how either person's wellbeing is of sufficient weight to place anyone under a duty. So the necessary condition set for a right by the Interest Theory is not satisfied. Call this the Problem of Harmless Wronging for the Interest Theory. The problem is that of accommodating our intuitions that Passenger and Target

---

[3] I omit that *Y*'s duty is *owed* to *X* as I see the Interest Theory as an account of *what it is* to owe a duty to another. Were we to allow "sufficient weight" to do a lot of heavy lifting, we could make Interest Theory (Canonical) necessary and sufficient for rights.

have their rights violated, given a commitment to the Interest Theory. More generally, the problem is that of accommodating agents having rights against harmless wrongs, given a commitment to the Interest Theory.

This chapter offers a principled solution to the Problem of Harmless Wronging by revising the Canonical statement of the theory with what I call the Safety Condition. The Safety Condition looks beyond what happens in the actual world to close possible worlds to normatively ensure that people's wellbeing is robustly protected across circumstances that could easily come about. In Sections 2 and 3, I introduce the Safety Condition by showing how it deals with preempted harm and pure risk imposition. By the end of Section 3, we see that preemption and pure risk differ from mundane cases of rights against harm in symmetric ways. Section 4 begins to offer an account of why rights might be sensitive to modality in the way the Safety Condition prescribes, before defending the Safety Condition against two objections.

A preliminary: suppose the Interest Theory fails. We should nonetheless want harm to explain Passenger's and Target's rights, even if we do not think *all* rights are grounded in their holder's wellbeing (and especially in *Hitmen*, introduced below). So those who deny the Interest Theory should still want a solution to the Problem of Harmless Wronging. Suitably refined, Safety Condition offers such an explanation.

2. Preemption

Let us begin by looking at the problem posed by our example of preempted harm, *Plane Crash*. An initial thought is that the problem posed by preemption is simply a problem with the Counterfactual Account of Harm. We need only refine or replace that. I think, though do not argue here, that we have good reason to hold onto the Counterfactual Account.[4] After introducing the Safety Condition, I return to this question of refining the Counterfactual Account.

Another tempting thought is that, although Passenger is not harmed all things considered by being denied admittance onto the plane, she is made worse off in a regard (she is harmed pro tanto). She suffers inconveniences that she would not have suffered were she not to have been arbitrarily denied

---

[4] For discussion, see, among others, Bradley (2012); Tadros (2016, 201–23).

admittance onto the plane.[5] Although the extent to which she is made worse off is insufficient to leave her harmed all things considered, there is none-theless *this regard* in which she is harmed. Perhaps she ought to be afforded a right protecting those aspects of her wellbeing. If we want to say this, we would need to endorse the following revision to the Interest Theory (Canonical):

Pro Tanto Thesis. X has a right that against Y that Y Φ, only if (and *because*) some aspect of X's wellbeing (an interest) is of sufficient weight to place Y under a duty to Φ.[6]

Because Passenger is made worse of in a regard, she can be afforded a right protecting those aspects of her wellbeing.

While the Pro Tanto Thesis might solve *Plane Crash* in that it accom-modates the intuition that Passenger has her rights violated, it will not solve all cases of preempted harm. It provides a solution to *Plane Crash* only because Passenger is made worse off along some specific dimension—a specific dimension that she would not have been made worse off in were Attendant to have allowed her onto the plane. However, this feature does not hold in all cases of preempted harm. Consider:

Hitmen. Suppose that we have two hitmen. $Hitman_2$ admires $Hitman_1$. $Hitman_2$ secretly follows $Hitman_1$ on every job she has in the hope that, one day, $Hitman_1$ will fail to complete a hit and she will be able to do so instead, thereby impressing $Hitman_1$.

For any victim that $Hitman_1$ is contracted to kill (call her Victim), what aspect of her wellbeing is set back by $Hitman_1$ that would not have been set back by $Hitman_2$? I am sceptical there is any such aspect, assuming $Hitman_2$ would have completed the hit the instant $Hitman_1$ fails to. Accordingly, the Pro Tanto Thesis cannot account for why Victim has a right against $Hitman_1$ killing her. We ought to look beyond the Pro Tanto Thesis.[7]

---

[5] We need to be careful not to introduce interests *created by* what we are owed, e.g. that Passenger has an interest in not being arbitrarily discriminated.

[6] e.g. Raz (1992, 129).

[7] Two notes. First, the way I am thinking about pro tanto harm is that we see if there is a regard that Victim is worse off in that she would not otherwise have been worse off in. Bradley thinks about pro tanto harm differently: 'Something is a [pro tanto] harm for a person if and

Returning to *Plane Crash*, the problem of preemption arises because we focus on a comparison between the following two worlds:

World 1. Attendant denies Passenger admittance onto the plane. The plane takes off and crashes, killing everybody on board.

World 2. Attendant does not deny Passenger admittance onto the plane. The plane takes off and crashes, killing everybody on board.

Passenger is no worse off in World 1 through Attendant's denying her admittance onto the plane than she would have been in World 2. However, World 2 is not the only possible world available for comparison with World 1. We might look to

World 3. Attendant does not deny Passenger admittance onto the plane. The plane takes off and lands safely at its destination.

Passenger is worse off in World 1 than she is in World 3. The extent to which she is worse off is sufficient to place Attendant under a duty not to deny her admittance onto the plane. World 3 is what we might call the *normal* counterfactual world. Perhaps we might revise the Interest Theory (Canonical) to respond to this fiat of the case.

Normality Thesis. X has a right against Y that Y Φ, only if (and *because*) X's wellbeing (her interests) is, *under normal circumstances*, of sufficient weight to place Y under a duty to Φ.

While the Normality Thesis does accommodate our intuition that Passenger has a right against Attendant, it will not solve all cases of preempted harm. For harm may be preempted in the normal world and yet we still want to say that people should be attributed rights against those preempted harms. Consider *Hitmen.* Because Hitman$_2$ *always* follows Hitman$_1$ on every job,

---

only if either (i) it is intrinsically bad for that person, or (ii) it brings about something intrinsically bad for that person, or (iii) it prevents something intrinsically good for that person' (2009, 66). If we suppose *Hitmen* involves painless death, only condition (iii) is relevant for our purposes, since death is only extrinsically good or bad. And since Hitman$_1$'s shooting Victim does not *prevent* anything intrinsically good for Victim, because Hitman$_2$ would have shot Victim anyway, condition (iii) is not satisfied. So Victim is not harmed pro tanto by Hitman$_1$ shooting her on Bradley's account, either. Second, a reader suggests in *Plane Crash* that Passenger is harmed by not being let onto the plane, though the short-term set back of her interests leads to a long-term improvement. But this analysis will not help with *Hitman.*

any victim that Hitman$_1$ is contracted to kill will not have a right against Hitman$_1$'s action. Victim would be no worse off under normal circumstances were Hitman$_1$ not to shoot, since Hitman$_2$ would make Victim worse off to an equal extent. The harm, then, is preempted under normal circumstances.

One might reply that it is not obvious that the circumstances described in this case are normal of hitmen in general, but only of Hitman$_1$. Perhaps we ought to restrict the reference class to *normal* hitmen. But this will not do either. Suppose that, so eager to make sure that their hits are completed, assassination agencies begin *always* to send their hitmen out in pairs. This means that for any victim it is normal that, were the first hitman to fail, another hitman would always be there to kill that victim.[8]

A less fanciful example than *Hitmen* is to imagine a polluting factory. Suppose these kinds of factories always pop up near each other, for example because they always require a river to dump waste in, cheap land, proximity to workforces, and so on. On the supposition that one polluting factory is sufficient to harm on its own, any harm caused by a particular factory would under normal circumstances be preempted by another factory. This is because there would always be another factory nearby, itself polluting, that would have caused an equal harm.[9]

Let us set out the problem with the Normality Thesis. We were faced with a comparison between two worlds:

Hitmen World 1. Hitman$_1$ fatally shoots Victim. (Hitman$_2$ was waiting in the wings.)

Hitmen World 2. Hitman$_1$ does not shoot Victim. Hitman$_2$ fatally shoots Victim.

Victim is no worse off in Hitmen World 1 than she is in Hitmen World 2. Hitmen World 2 is what would happen under normal circumstances were Hitman$_1$ not to have shot. However, as with *Plane Crash*, there is another close world that we can appeal to:

---

[8] This holds on most accounts of normality, including non-statistical ones, e.g. Smith (2016, 38–45).

[9] "Cancer Alley" comes to mind, an 85-mile stretch of the Mississippi that produces one quarter of America's petrochemicals.

Hitmen World 3. Hitman$_1$ does not shoot Victim. Hitman$_2$ does not shoot Victim.

Victim is worse off in Hitmen World 1 than she is in Hitmen World 3. The question is how we make reference to *these* close worlds (Plane Crash World 3 and Hitmen World 3) to accommodate the intuitive verdict that Passenger and Victim have rights against the relevant conduct.

I suggest we make an appeal to *modal safety*. The idea behind safety is nicely explained by Timothy Williamson:

> Imagine a ball at the bottom of a hole, and another balanced on the tip of a cone. Both are in equilibrium, but the equilibrium is stable in the former case, unstable in the latter. A slight breath of wind would blow the second ball off; the first ball is harder to shift. The second ball is in danger of falling; the first ball is safe. Although neither ball did in fact fall, the second could easily have fallen; the first could not.   (Williamson 2000, 123)

There is a danger an event will occur if that event does occur in some sufficiently similar case. In much the same way as the ball is not safely balanced on the top of the cone, Passenger's and Victim's wellbeing is not safely protected—though they are not actually made worse off, there is a danger they could have been. And, it is plausible that rights ought to safely protect their wellbeing.

We can make appeal to safety by revising the Interest Theory (Canonical):

Interest Theory (Safety). X has a right against Y that Y Φ, only if (and *because*) Y's not Φ-ing causes X to be worse off than she would have been in at least one close world, and the difference in X's wellbeing is of sufficient weight to place Y under a duty to Φ.

For brevity, I call this the Safety Condition. It works by comparing how X fares when Y does not act as the would-be duty dictates with how X would have fared in close worlds in which Y acts as the would-be duty dictates. Call the world in which Y does not act as the would-be duty dictates our *world of evaluation.* Call the world in which Y does act as the duty dictates our *world of comparison.* (This jargon becomes useful in the following section.) In *Plane Crash* and *Hitmen*, the world of evaluation is World 1, the actual world. We evaluate how our potential right-holder fares in this world through comparison with close worlds in which the potential duty-bearer acts as the would-be duty dictates (hence the names: world of evaluation and of comparison). Since Passenger and Victim are worse off in World 1 than they are in World 3, and since the extent to which they are worse off is of sufficient weight to place Attendant and Hitman$_1$ under their respective duties, Passenger and Victim hold rights. They hold these rights in the actual world.

As with the Interest Theory (Canonical), the Safety Condition is only a necessary condition.[10] So, it does not follow that Passenger and Victim hold rights just because the Safety Condition is satisfied. But we can assume the other conditions necessary and jointly sufficient for rights are satisfied. We turn to these other conditions in Subsection 4.3.

Above, I questioned whether we ought not refine the Counterfactual Account of Harm. One might wonder whether we ought to endorse the Safety Condition as an account of harm rather than as a refinement to the Interest Theory.[11] Extensionally, there is nothing between building safety into our account of rights or our account of harm. I would not be too worried if one takes all I say about the Safety Condition and builds it into their account of harm. However, let me offer two reasons why I prefer building it into our account of rights.

First, I do not believe that merely being subjected to risk is itself harmful (Bowen Forthcoming). As we see in the following section, the Safety Condition is satisfied in cases of pure risk, such as *Roulette*. If we build the Safety Condition into our account of harm, this means that merely being subjected to risk is itself harmful. So, it is preferable to build the Safety Condition into our account of rights—and say, though Shooter does not harm Target, she does violate her rights—rather than our account of harm.

Second, as we also see below, it appears that the Safety Condition over-generates. If we build safety into our account of harm (and benefit), it overgenerates harms (and benefits). If we build safety into our account of rights, it overgenerates rights. I argue we can explain away these over-generations. However, the sorts of considerations I appeal to in order to explain away these overgenerated rights are not the sort of features that

---

[10] Again, we could allow "sufficiently weighty" to do some heavy lifting, and make the Safety Condition necessary and sufficient.

[11] This is something like Tadros's *Complex Counterfactual View of Harm* (2016, 177).

ought to affect whether or not one has *harmed* another.[12] So, it is better to build the Safety Condition into our account of rights.

3. Risk

We have seen how the Safety Condition deals with cases of preempted harm. In this section, we see how the Safety Condition deals with cases of pure risk imposition.

In *Roulette*, Shooter comes into Target's room and plays a round of Russian roulette with her. Since no bullet is fired, Target is no worse off than she would have been had Shooter not subjected her to that risk. Because of this, Target's wellbeing is not protected by her right not to be subjected to gratuitous risk of harm. Given the Interest Theory (Canonical), Target has no right against Shooter that Shooter not play roulette with her.

One might hold Target has a right against Shooter because, given the best available evidence to Shooter, in expected terms Target's wellbeing is of sufficient weight to place her under a duty not to play Russian roulette with Target. On this view, rights are determined from the evidence-relative perspective, rather than the fact-relative perspective as I have been assuming.[13] I think there are problems with Evidence-Relative Views. For example, these views imply that what rights I hold *change* as new evidence becomes available to others, rather than others getting better evidence about what rights exist. These views also imply, if the best available evidence does not support that someone's behaviour will harm me (or, for other sorts of reasons, does not support that I have a right against being harmed), that I have no right against that harmful behaviour. And there are lots of cases in which this looks very implausible (Bowen 2021). I do not say more about Evidence-Relative Views here. Instead, notice that even those very confident in Evidence-Relative Views need to see how they compare to the best alternative views. We can see the Safety Condition as offering an alternative Fact-Relative View of rights.

The Safety Condition can accommodate this verdict that Target has a right against Shooter that Shooter not subject her to such risky behaviour.

---

[12] An example, preempting Subsection 4.3: while we may think one's intentions or knowledge is relevant to whether others hold rights against one, it ought not affect whether one *harms* others.

[13] e.g. Zimmerman (2014); Quong (2015); Vossen (2016); Quong (2020). Quong's view is actually something of a hybrid. For discussion, see my Bowen (2021).

Showing this requires looking back to preemption. Like Passenger in *Plane Crash*, Target is no worse off than she would have been had the duty correlative to her right been respected. But, unlike in cases of preemption, in which another event causes her to be at least as badly off as she would have been had the duty not been respected, this is because the risked harm does not materialize. We are comparing, then:

Roulette World 1. Shooter plays Russian Roulette. No bullet is fired.

Roulette World 3. Shooter does not play Russian Roulette.

Target is no worse off in Roulette World 1 than she is in Roulette World 3. However, as with preemption, there is an extremely close world against which Target is made worse off because of Shooter playing roulette with her—namely, the world in which there *is* a bullet in the chamber when Shooter pulls the trigger:

Roulette World 2. Shooter plays Russian Roulette. Shooter fatally shoots Target.

Unlike when dealing with preemption, in which we compare World 1 with World 3, in cases of pure risk imposition we compare World 2 with World 3. Target is worse off in Roulette World 2 than she is in Roulette World 3. The extent to which she is worse off is sufficient to place Shooter under a duty not to make Target the subject of her risky behaviour. The Safety Condition can accommodate this verdict. It requires recognizing Roulette World 2 as our focus when Y fails to act as the duty dictates instead of the actual world, Roulette World 1—it requires recognizing Roulette World 2 as the *world of evaluation* rather than the actual world, Roulette World 1.

So, in cases of preemption, our focus when Y does not act as the duty dictates (our world of evaluation) is World 1, the actual world. In cases of pure risk imposition, our focus is World 2 (a world close to the actual world). It is in keeping with the Safety Condition for our focus on what happens when Y fails to act as the duty dictates to include worlds close to the actual world, rather than only the actual world itself. This is because we could easily have been in those close worlds—it could easily have been that there was a bullet in the chamber when Shooter pulls the trigger. While different events occur between Roulette World 1 and World 2 (Victim is not and is shot), Shooter *acts*, in the relevant sense, in the same way in both worlds. And this is consistent with the idea behind the Safety Condition: we are unsafe to the extent that our wellbeing is not safely protected; as Shooter could easily have shot Target, Target's wellbeing was not safely protected; by allowing us to focus on worlds in which Shooter acts in the same way close to the actual world, rather than only the actual world, Target's wellbeing is safely protected. So far, so good.

At this stage, one might wonder how we determine which world to focus on as our world of evaluation when Y does not act as the duty dictates. Why focus on the actual world in cases of preemption but focus on worlds close to the actual world in cases of pure risk? But we do not need to determine which of these worlds to uniquely focus on. While the Safety Condition speaks of *the* world of evaluation, this need not imply that there is only one world of evaluation. Rather, there are multiple worlds of evaluation—those close worlds in which the potential duty-bearer performs the act that she may be under a duty not to perform. (This includes the actual world, for the actual world is close to itself.)[14] In *Roulette*, worlds of evaluation are close worlds in which Shooter puts a bullet in the cylinder of the gun, spins the cylinder, and pulls the trigger. In *Plane Crash*, worlds of evaluation are close worlds in which Attendant denies Passenger admittance onto the plane. In *Hitmen*, worlds of evaluation are close worlds in which Hitman$_1$ shoots Victim. Again, this is in line with the Safety Condition: it ensures Target's wellbeing is robustly protected. We do not need to determine which world to focus on when Y does not act as the duty dictates since the Safety Condition looks to *all* close worlds in which the duty-bearer acts in the same way.

With this in place, we can better appreciate how the Safety Condition works. It makes for a two-part comparison. We have worlds of evaluation and worlds of comparison. Worlds of evaluation are worlds in which the potential correlative duty-bearer performs the action that she may be under a duty not to perform. Worlds of comparison are worlds in which the potential correlative duty-bearer does *not* perform this action. Whereas the Canonical statement of the Interest Theory focuses only on the closest worlds in which the duty is and is not respected, the Safety Condition looks also to other worlds that are (sufficiently) close to the closest worlds in which the duty is and is not respected. And if the potential right-holder is worse off in a world of evaluation than she is in a world of comparison, and the extent

---

[14] While this is true on most accounts of closeness, it is not necessarily true, e.g. Smith (2016).

to which she is worse off is sufficient to place the potential correlative duty-bearer under a duty, the Safety Condition is satisfied.

As a point of comparison, consider a mundane case in which one's rights against harm are violated: suppose Threatener punches Innocent. Here, we compare the actual world, in which Threatener punches Innocent, with the closest counterfactual world, in which Threatener does not punch Innocent. Our world of evaluation is the closest world (to the actual world) in which Threatener punches Innocent: the actual world. Our world of comparison is the closest world (to the world in which Threatener punches Innocent) in which Threatener does not punch Innocent. Since Innocent is sufficiently worse off in the world in which Threatener punches her than in the world in which Threatener does not punch her, Threatener is under a duty not to punch Innocent correlative to Innocent having a right that Threatener not punch her.

In cases of preemption, we compare how our potential right-holder fares not with how she fares in the closest counterfactual world in which the potential duty is respected, but with a different counterfactual world. In cases of pure risk, we do use the closest counterfactual world in which the potential duty is respected. However, we do not use the actual world as our world of evaluation, but some world close to the actual world. To this extent, preemption and pure risk differ from the mundane case in symmetrical ways. They are cases of harmless wronging in symmetrical ways to each other. (We could also have a case of preempted pure risk.)

To be precise when expounding the Safety Condition, I have used some jargon. Here is a more natural gloss on the Safety Condition. For someone to hold a right against us that we do not perform some action, we look to whether our performing that action could easily leave them sufficiently worse off to place us under a duty.

4. In Defence of Safety

Whereas the standard version of the Interest Theory and refinements considered above fall foul of the Problem of Harmless Wronging, the Safety Condition correctly generates rights in our cases of preemption and pure risk imposition. Further, the Safety Condition offers us a unified account of *why* people are attributed rights across these different types of case: even if one is not made worse off as things turn out, one could easily have been made worse off. The Safety Condition's *principled* extensional accuracy is the primary virtue of the account that I would like to stress in this chapter.

In this section, I defend the Safety Condition against two objections: that it under- and overgenerates rights. Before that, I say a little about *why* modality might matter for rights in the way that the Safety Condition prescribes.

4.1 Why Safety?

One important feature of the Safety Condition is that it removes an objectionable form of luck from rights. If we assume the Interest Theory (Canonical) then, through sheer luck, Attendant does not violate Passenger's right not to be denied admittance onto the plane. It could easily have been that Attendant *did* harm Passenger, and so would have violated her rights. Similarly, through sheer luck Shooter does not violate Target's rights. It could easily have been that there was a bullet in the chamber when Shooter pulled the trigger, and so Shooter would have violated Target's rights. By focusing on more than what happens in the actual world, the Safety Condition removes this objectionable form of luck from rights (and right-violations). This gives us good reason to endorse the modal character of the Safety Condition.

One might wonder why it matters that rights do not depend on (objectionable) forms of luck. Some hold that others are worse off to the extent that their wellbeing depends on luck (Lazar 2017; Oberdiek 2017). This is usually couched in terms of luck being antithetical to control, and control being necessary for one to lead an autonomous life. If we were to go this way, it would leave the Safety Condition unmotivated as regards those for whom autonomy has little or no value (for example, very young children and those with severely damaged rational capacities) (Bowen Forthcoming). And, regardless, I believe that at least rights against *harm* should be grounded in facts about wellbeing in general, rather than autonomy.

Instead, I propose it matters that rights do not depend on luck in this objectionable way—or, that rights depend on modality as the Safety Condition prescribes—because this formally requires that we are sensitive to others' wellbeing: that it is not merely that we do not harm others, but that we could not easily have harmed others.[15]

The Interest Theory starts with the idea that others' wellbeing is sufficiently important to place others under duties. You need not be an Interest Theorist to believe that. Interest Theorists are distinctive because they say, a nice explanation for when and why you owe someone a duty is that their wellbeing is the grounds of the duty. Commonly, it is taken that duties are just a special type of reason: they have something like exclusionary weight, they leave a moral remainder when not acted on, they are demandable, and so on. The Safety Condition's focus on what could easily happen makes duty-bearers' reasons of this kind more sensitive to others' wellbeing than is the case on the modally undemanding, Canonical statement of the Interest Theory. Since the Interest Theory began with the idea that others' wellbeing is very important—it both places us under duties and exclusively makes those duties owed to others—it is in keeping with the Interest Theorists' motivations to endorse the Safety Condition.

Think back to Williamson's example above. Suppose you have asked two people to put the ball somewhere and keep it still. When all else is equal, the person who puts the ball at the bottom of the hole has taken more care to ensure that the ball is still than the person who balanced the ball on the top of the cone. Analogously, the person who turns out not to harm others, but could easily have, such as Attendant, has taken less care not to harm others than those who robustly do not harm others. This offers an attractive picture of what we *owe* to others—that we take care not to leave others worse off through our actions than otherwise could have been the case.

This is just the beginning of an account of *why* rights respond to modality in the way the Safety Condition prescribes. But it is helpful to have more than extensional accuracy behind the Safety Condition before turning to our objections.

(One question this prompts: does it matter that we are sensitive to others' wellbeing only when we are uncertain of how our actions will turn out? When harm is preempted, I am tempted to say that we have rights against others' actions even if they know their harming us is preempted. This

---

[15] In saying that duty-bearers are sensitive to others' wellbeing, one might think I am confusing my modal notions. Y's Φ-ing *safely* does not harm X iff there is no close world in which Y's Φ-ing harms X. Y is *sensitive* to her Φ-ing not harming X iff, were Φ-ing harmful, Y would not Φ. That Y safely does not harm X does not mean that she is sensitive to her Φ-ing not harming X. To be more precise, I should say the Safety Condition makes Y sensitive to her Φ-ing not being harmful in sufficiently close worlds.

explains why, even if someone's contribution to my being harmed is pre-empted, I could use them as a means to prevent myself from being harmed. I am less sure when it comes to pure risk. How much would I pay not to be subjected to a merely modal risk?[16] Little, if anything. This suggests possible worlds do not matter when we know they are not going to come about. On this way of thinking about things, while evidence affects which possible worlds are relevant to rights, it is very far from Evidence-Relative Views mentioned in Section 3. On this version of the Safety Condition, rights depend only on the facts—it is just that modal facts about what could otherwise have been the case might lose their relevance if we *know* those worlds are merely modal.)

4.2 Very Preempted Harm

In *Plane Crash*, the Safety Condition works by comparing the actual world with the close world in which Passenger boards the plane and it does not crash. In *Hitmen*, the Safety Condition works by comparing the actual world in which $Hitman_1$ shoots Victim with the close world in which $Hitman_1$ does not shoot Victim and $Hitman_2$ does not shoot Victim. In both cases, we might say, we are looking past the *closest* (counterfactual) world to some other close (counterfactual) world in which the preempting harm does not occur. However, here lies the recipe for a counterexample: make the harm *very* preempted:

Smokin' Aces. $Hitman_1$ is contracted to kill Victim. Unbeknown to $Hitman_1$, there are one hundred other hitmen waiting in the wings. Each is ready to kill Victim if the previous hitman fails.

In *Smokin' Aces*, the Safety Condition has to compare

(*Smokin' World 1*) $Hitman_1$ shoots Victim ($Hitmen_{2-100}$ were waiting in the wings),

with

---

[16] Thanks to Peter Graham for putting it like this.

(*Smokin' World 101*) Hitman$_1$–Hitman$_{100}$ all do not shoot Victim.

Victim is worse off in Smokin' World 1 than she is in Smokin' World 101—Smokin' World 101 is the closest world that allows us to say this. However, one might hold that Smokin' World 101 is *not* a close enough world for the purposes of satisfying the Safety Condition. So, Victim will not have a right that Hitman$_1$ not shoot her.

Why is Smokin' World 101 is not a close world? Consider David Lewis's view of closeness. Begin by supposing that the worlds under consideration have deterministic laws: they start and play out following deterministic laws of nature. We discover the closeness of two worlds at some time, t, by reference to the number and size of the violations of those laws of nature that would be required at t to render those worlds convergent after t (Lewis 1979, 472). We need 100 miracles to get us from Smokin' World 101 to Smokin' World 1. And perhaps that is too many miracles.

One way to address this Problem of Very Preempted Harm is to lean on our theory of closeness to avoid the verdict that the world in which the harm is not preempted is not close. For example, because the differences between Smokin' World 1 through to Smokin' World 101 are *themselves* so similar, one could argue Smokin' World 101 is *not* actually all that far from Smokin' World 1. Call this the *Similarity Strategy*. Taking this strategy, we have not relaxed how close a world needs to be for it to be close enough to satisfy the Safety Condition (which would make the view more likely to overgenerate rights), but suggested that Smokin' World 101 is not actually that far away, so will satisfy the Safety Condition. We might even say, how far away these worlds are tends towards some limit, no matter how many hitmen we add.

There is at least one limitation with the Similarity Strategy. It works with *Smokin' Aces* because the many different preempting acts are relevantly similar, thereby diminishing their impact on how far away worlds are. But maybe this is just a fiat of *Smokin' Aces*. Perhaps one could devise a case in which all the different things that cause the harm to be preempted are of a different character, meaning that the Similarity Strategy will be of little help. We could suppose that were Hitman$_1$ not to have killed Victim, a boulder would have fallen on her and, had the boulder not fallen, lightning would have struck her, and so on.

Another way to resist the Problem of Very Preempted Harm is to revise the Safety Condition. The element that cases of very preempted harm put pressure on is that X must be worse off in at least one *close* world. We could revise this closeness element:

Interest Theory (Safety, Relevance Variant). X has a right against Y that $Y \Phi$, only if (and *because*) Y's not Φ-ing causes X to be worse off than she would have been in at least one relevant world, and the difference in X's wellbeing must be of sufficient weight to place Y under a duty to Φ.

Since there is no stipulation that the world of comparison needs to be a close world, we need not worry that Smokin' World 101 is far away from Smokin' World 1. It turns only on whether we think Smokin' World 101 is relevant. And it seems relevant.

There is an obvious worry with the Relevance Variant: how do we work out which worlds are relevant? Despite the Relevance Variant faring better with the Problem of Very Preempted Harm than the standard Safety Condition, I am not sure that there is a non-circular way to answer this question. So while the Relevance Variant may score well extensionally, it has less explanatory power than the Safety Condition.

But perhaps this worry is premature. Speaking of the safety condition on knowledge, Williamson says: 'In many cases, someone with no idea of what knowledge is would be unable to determine whether safety obtained... One may have to decide whether safety obtains by first deciding whether know-ledge obtains, rather than vice versa' (2009, 305). On this way of under-standing things, safety can be thought of as offering a circular account of knowledge. Return now to the Relevance Variant of the Safety Condition. Even if we cannot know whether one is owed a duty and holds a correlative right by the lights of the Relevance Variant alone, that does not mean that it is ontologically circular. We can still maintain that whenever one is owed a duty, and holds a right correlative to that duty, the duty is grounded in how one fares across relevant worlds. This is so even if we need a prior under-standing of whether one is owed that duty to know whether the world is as relevant.

There is a third response to Very Preempted Harm, one that does not require moving to the Relevance Variant: we might accept our problem's conclusion. When harm is preempted to a great degree, along many differ-ent dimensions, perhaps people do not have rights against that harm. Let me offer two remarks to make this bullet easier to swallow.

First, above we considered only one feature of closeness (the Similarity Strategy) that helps solve the problem. We might hold out hope that there are other features of the semantics of closeness that will provide a solution to our problem.

Second, and more substantially, we are attempting to offer a *reductive* account of rights (and directed duties)—an account that explains rights and what it is to owe a duty to another person by appealing to some other feature(s). The Interest Theory (with the Counterfactual Account) began with the idea that rights are difference makers. The Safety Condition adds, "rights are difference makers or *could-easily-have-been difference makers.*" The Problem of Very Preempted Harm presses, "Well, what if the putative duty-bearer couldn't easily have been a difference maker?" Perhaps it should not be surprising that we need to accept some counterintuitiveness along this line. When someone really is doomed to suffer some bad fate, perhaps they do not have rights against us that we not make that fate preempted to more of an extent.[17]

4.3 Overgenerations

Whereas the Problem of Very Preempted Harm suggests that the Safety Condition will undergenerate rights, the Safety Condition may also over-generate rights. The Safety Condition requires only that there is one close world in which the right-holder fares sufficiently better through the duty-bearer acting as the duty requires. But there are many close worlds. Will not the Safety Condition be too easily satisfied?

Take *Plane Crash*. There is a close world, through comparison with which Passenger is made worse off by being denied admittance onto the plane. The Safety Condition is satisfied, so Passenger has a right against being denied admittance onto the plane. However, there is also another close world, through comparison with which Passenger is made *better off* by being denied admittance onto the plane: through comparison of World 1 and World 2. Does this imply that Passenger has a right against Attendant that Attendant deny Passenger admittance onto the plane? It might appear so— the Safety Condition *has* been satisfied. This is one example of the Problem of Overgeneration.

Consider another case:

---

[17] Most difficult to swallow are going to be cases in which people work together to make harm very preempted. But perhaps there are things to say in reply to these cases—for example, that the victim has a right against the group acting.

A&E. An unconscious Patient comes into A&E with a burst aneurysm. There is no chance that the aneurysm will stop bleeding spontaneously and so, without treatment, Patient will surely die. Surgeon can operate. The surgery is very serious, though all-things-considered beneficial.

While the case stipulates that there is no nomologically possible world in which Patient recovers without treatment (a possible world with our laws of nature), there might be a metaphysically close world in which the aneurysm stops bleeding. On the Lewisian view, all that would be required is a small miracle to stop the aneurysm from bleeding. Through comparison with this world, Patient would be worse off as a result of the operation: she will unnecessarily have gone through a very serious operation. Does this imply that Patient has a right against Surgeon performing the operation *because* there is a metaphysically close, though nomologically impossible, world in which the bleeding would stop?

I think the Problem of Overgeneration can be resisted. In what follows, I do so on several fronts. The general strategy is that we rule out worlds that would otherwise allow the Safety Condition to overgenerate rights by appealing to other features of closeness or to other necessary but insufficient conditions on rights.

Before that, note the Problem of Overgeneration does not arise if we prefer the Relevance Variant of the Safety Condition. Since there is no stipulation that close worlds are relevant, we need not worry that World 2 is very close in *Plane Crash*, since we can say World 2 is not relevant. Similarly, we need not worry about the metaphysically close world in which Patient's burst aneurysm stops bleeding. Obviously enough, one might still be worried about the view's epistemic circularity. So, let us see what can be said on behalf of the Safety Condition without moving to the Relevance Variant.

First, we may introduce what we can call the *Realism Condition*, according to which only those worlds that are nomologically possible count as close for rights.[18] In *A&E*, the metaphysically possible world in which the aneurysm stops bleeding will not be close, so the Safety Condition will not be satisfied. However, the Realism Condition takes us only so far. It is of little use in *Plane Crash*.

---

[18] See McMahan (2002, 133–6).

Second, another feature of closeness that helps us deal with cases like *Plane Crash* is that the closeness relation is what we can call *nonreciprocal*—just because w_1 is close to w_2 when w_1 is our focus does not entail that w_2 is close to w_1 when w_2 is our focus.[19] For example, though if the plane crashes there *is* a close world in which the plane does not crash, it is often *not* the case that, if the plane does not crash, there is a close world in which the plane does crash. Roughly, this is because a lot needs to go wrong for a plane *to* crash—meaning the world in which it does crash is far away. But it does not take a lot for a plane *not* to crash—meaning that the world in which the plane does not crash is close. While this does not explain why Passenger does not have a right that *she* be denied admittance onto the plane (since in that case, there is a close world in which the plane crashes), it does explain why, generally, *we* do not have rights to be denied admittance onto planes—usually, there is no close world in which the plane crashes. The Problem of Overgeneration is not as worrying as it might have seemed.

Third, recall the Safety Condition begins:

X has a right against Y that Y Φ, only if (and *because*) Y's not Φ-ing *causes* X to be worse off...

That Y needs to *cause* X to be worse off seems to get the cases right: Attendant causes Passenger to be worse off through being denied admittance onto the plane were the plane not to have crashed; Hitman$_1$ causes Victim to be worse off were Hitman$_2$ not present; and Shooter causes Target to be worse off were the risk to materialize. But, were Attendant to allow Passenger onto the plane, she would not *cause* Passenger to be worse off.[20] Given this, the Safety Condition is not satisfied, so Passenger will not have a right that Attendant deny her admittance onto the plane.

There may be problems with this causal restriction on the relationship between Y's Φ-ing and how X fares, as there might be cases in which Y's not Φ-ing does not cause X to be worse off, yet we think X has a right against Y that she Φ. Without a theory of causation on the table, we are not able to fully assess this issue. Instead, let us move onto one final way to resist the Problem of Overgeneration: the Safety Condition is a necessary but

---

[19] For similar discussion, see Lewis (1973, 50–2).
[20] This is not because omissions are not causal (Lewis 1986, 189–93; Paul and Hall 2013, 173–214).

insufficient condition on rights. And this means that X might not have a right that Y Φ even if there is a close world in which Y's Φ-ing leaves X much better off.

So, what other necessary but insufficient conditions would need to be satisfied? Consider the following two popular moral asymmetries:

Doctrine of Double Effect (DDE). Other things being equal, it is harder to justify Y intending to harm X than it is to justify Y merely foreseeing that she will harm X.

Doctrine of Doing and Allowing (DDA). Other things being equal, it is harder to justify Y doing harm to X than it is to justify Y allowing harm to X.

Suppose we made DDE internal to rights. We would say that, all else equal, it is more likely to warrant a right if Y intends the harm that would befall X were she to violate the putative duty than if that harm were merely foreseen. This means that a harm to X caused by Y might be sufficient to warrant a right if that harm is intended by Y, but insufficient if it was an unintended.

While both DDE and DDA are *ceteris paribus* conditions, and stated generally, endorsing either (or both) helps resist the Problem of Overgeneration. Suppose Attendant were to admit Passenger onto the plane and the plane were to crash. Given the way that the case is stipulated, it is unlikely that Attendant *intends* the harm or *does* harm to Passenger. This helps explain why Passenger fails to have a right that Attendant deny her admittance onto the plane (notwithstanding my suggestion above that the Safety Condition is not satisfied in this variant in any case, because Attendant would not *cause* harm to Passenger were she to allow her onto the plane).

I have offered three dimensions along which Passenger might fail to have a right that Attendant deny her admittance onto the plane. First, Attendant would not cause Passenger to be worse off were she to allow Passenger onto the plane. Second and third, Attendant does not intend nor *do* the harm to Passenger. Taken with the two considerations above—that only nomologically worlds count as close for the Safety Condition and that the closeness relation is nonreciprocal—a picture is emerging of how we might respond to the Problem of Overgeneration.

5. Conclusion

I have introduced the Safety Condition as a reply to the Problem of Harmless Wronging for the Interest Theory. The problem is that of accommodating our intuitions that agents have rights against harmless wrongs, given a commitment to the Interest Theory. The Safety Condition's extensional accuracy and unified account of *why* agents are attributed rights across these different types of case of harmless wronging are the primary virtues of the account that I have stressed in this chapter.

6. Acknowledgements

This chapter is based on my PhD thesis, so there are more people to thank by name than space permits—but thanks to everyone with whom I have discussed safety, and especially to those who have provided me with written comments on drafts of this chapter, including two anonymous readers for this volume. Thanks to audiences in Cambridge, Edinburgh, Oxford, St Andrews, Boulder (and especially to Bodhi Melnitzer, my excellent commentator), Berlin, and to those who attended two online workshops at which this chapter was presented. Finally, thanks to my supervisors, Rowan Cruft and Theron Pummer, and to Stephanie Bowen, for their unfaltering support. This work was supported by an AHRC Scottish Graduate School for the Arts and Humanities doctoral training partnership studentship (grant no. 1804818) and a Society for Applied Philosophy Doctoral Scholarship.

Works Cited

Bowen, Joseph. 2021. 'Review of Jonathan Quong, The Morality of Defensive Force'. *Ethics* 131 (3): 625–30.

Bowen, Joseph. Forthcoming. '"But You Could've Hurt Me!": Risk and Harm'. *Law and Philosophy.*

Bradley, Ben. 2009. *Well-Being and Death.* Oxford University Press.

Bradley, Ben. 2012. 'Doing Away with Harm'. *Philosophy and Phenomenological Research* 85 (2): 390–412.

Finkelstein, Claire. 2003. 'Is Risk a Harm?' *University of Pennsylvania Law Review* 151 (3): 963–1001.

Kamm, F. M. 2007. *Intricate Ethics.* Oxford: Oxford University Press.

Lazar, Seth. 2017. 'Risky Killing: How Risks Worsen Violations of Objective Rights'. *Journal of Moral Philosophy* 16 (1).

Lewis, David K. 1973. *Counterfactuals*. Oxford: Basil Blackwell.

Lewis, David K. 1979. 'Counterfactual Dependence and Time's Arrow'. *Noûs* 13 (4): 455–76.

Lewis, David K. 1986. 'Causation'. In *Philosophical Papers. Vol. 2*, 159–72. Oxford: Oxford University Press.

McMahan, Jeff. 2002. *The Ethics of Killing: Problems at the Margins of Life*. Oxford: Oxford University Press.

Oberdiek, John. 2017. *Imposing Risk: A Normative Framework*. Oxford: Oxford University Press.

Paul, L. A., and Ned Hall. 2013. *Causation: A User's Guide*. Oxford, New York: Oxford University Press.

Quong, Jonathan. 2015. 'Rights Against Harm'. *Aristotelian Society Supplementary Volume* 89 (1): 249–66.

Quong, Jonathan. 2020. *The Morality of Defensive Force*. Oxford: Oxford University Press.

Raz, Joseph. 1992. 'Rights and Individual Well-Being'. *Ratio Juris* 5 (2): 127–42.

Smith, Martin. 2016. *Between Probability and Certainty: What Justifies Belief*. Oxford: Oxford University Press.

Tadros, Victor. 2016. *Wrongs and Crimes*. Oxford: Oxford University Press.

Thomson, Judith Jarvis. 1990. *The Realm of Rights*. Cambridge, Mass.: Harvard University Press.

Vossen, Bas van der. 2016. 'Uncertain Rights Against Defense'. *Social Philosophy and Policy* 32 (02): 129–45.

Williamson, Timothy. 2000. *Knowledge and Its Limits*. Oxford: Oxford University Press.

Williamson, Timothy. 2009. 'Replies to Critics'. In *Williamson on Knowledge*. Oxford: Oxford University Press.

Zimmerman, Michael J. 2014. *Ignorance and Moral Obligation*. Oxford: Oxford University Press.

3

Solidarity in Kantian Moral Theory

Adam Cureton

Critics of Kantian moral theory worry that it denies, downplays, or misrepresents the role of friendship, community, and other loving relationships in the moral life.[1] My aim is to describe a new and unorthodox way for such theories to incorporate and justify the status of certain relationships as permissible, obligation-generating, and worthy of promotion, protection, and respect.[2] Contrary to what most defenders and critics of Kant usually believe, I argue that one theme in his thinking is that human persons have intrinsic rational interests in forming, maintaining, perfecting, respecting, and promoting relationships of solidarity.[3] Part of being a rational, or as we might say, a rational and reasonable, person is to be concerned with relationships of this sort apart from any natural desires and feelings we might have.

Attributing these "interests of reason,"[4] which are ends, drives, concerns, dispositions, or needs of reason itself, to human persons provides us with rational grounds for what we could rationally will as universal law or for what we would agree to as rational legislators in a kingdom of ends. Understood as a principle of justifiability, the Categorical Imperative requires us to conform to laws that are rationally justifiable to everyone,

---

[1] Williams (1981), Annas (1984), Wolf (1992, 2012), McDowell (1994), Held (2006, 97–9).

[2] I am grateful to Tom Hill, Mark Timmons, and audiences at the 2021 WINE conference and the 2021 Central APA for their feedback on this essay.

[3] I do not claim that this theme is part of Kant's considered philosophical theory or the best interpretation or rational reconstruction of it; nor do I take a position on its status, relative priority, and relationship to other themes in Kant's thinking.

[4] e.g., G 4:460n; MM 6:212–13; A462/B490-A476/B504; A741/B769; CPJ 5:223; NF 18:274 (see n. 5 for a key to abbreviations). In some places, Thomas E. Hill, Jr., gestures towards a more expansive set of rational interests that help to determine how his ideal legislators deliberate and legislate (Hill 2000, 139, 150–1, 2002b, 152–3). Rawls (1999, 312–13) also claims that persons have intrinsic rational interests in promoting and protecting their rational powers to form a conception of the good and to act from principles of justice. A few other Kantians, including Yovel (1989), Kleingeld (1998), Velkley (2014), Ferrarin (2015, 24–34), and Raedler (2015, 12–15, 60–66), have noticed and discussed Kant's claims about reason's interests.

Adam Cureton, *Solidarity in Kantian Moral Theory* In: *Oxford Studies in Normative Ethics, Volume 12.*
Edited by: Mark Timmons, Oxford University Press. © Adam Cureton 2022.
DOI: 10.1093/oso/9780192868886.003.0004

but many of its formulations share a common need to specify criteria that determine whether putatively rational laws are rationally justifiable to persons. Our rational interests in solidarity provide some of these criteria. By analogy, much as a rational being "cannot possibly will" a universal law of nature not to develop his natural talents because "as a rational being he necessarily wills that all his capacities in him be developed," rational people could not, for example, agree to a universal law that prevents them from having relationships of solidarity with others because such a law conflicts with their rational interest in being part of such relationships.[5]

My plan for this chapter is as follows. First, I highlight several moral dimensions that, in common sense, loving relationships seem to have. Second, I explore the charge that Kantian moral theories cannot adequately capture and explain these ordinary moral judgments and argue that this objection is graver than many of Kant's detractors and defenders seem to think. Third, I describe an underappreciated theme in Kant's thinking, which appears in his main published works as well as his other writings and lecture notes, namely that the faculty of reason in each of us, as it were, has its own substantive and intrinsic interests that help to determine what moral principles it legislates. Fourth, I draw on and abstract from Kant's discussions of various relationships to characterize a kind of solidarity. Fifth, I provide some reasons for thinking that, according to Kant, each of us has rational interests in establishing relationships of solidarity with other people, maintaining and perfecting ones we are in, promoting such relationships among others, and respecting relationships of solidarity themselves. Sixth, I consider what specific kinds of *prima facie* moral principles might be justifiable to all rational people in light of their rational interests in solidarity. I end by briefly noting how Kant's conception of a morally perfect world or kingdom of ends includes and partially consists of many kinds of solidary relationships.

---

[5] G 4:423. I will refer to Kant's works with these abbreviations followed by standard Academy volume and page numbers: A,B—Kant (1998), Anth—Kant (2007a), CPJ—Kant (2000), CPrR—Kant (2007b), Eth-C—Kant (2001c), Eth-H—Kant (2001b), Eth-V—Kant (2001a), G—Kant (1996a), L-Anth—(Kant 2013), L-Log—(Kant 1992), MM—Kant (1996b), NF—Kant (2005), Ped—Kant (2007c), Rel—Kant (2001d), TP—Kant (1999a), TPP—Kant (1999b), WOT—Kant (2001e).

I. Moral Dimensions of Loving Relationships

Let's begin with some examples that illustrate several kinds of apparently moral dimensions of loving relationships.[6]

A. Forming Loving Relationships

David lives alone on a secluded mountaintop where he has no friends, family, spouse, or other personal relationships; he mainly sees his limited interactions with others as ones of convenience or utility; and he is, by his own lights, quite content and well-off in his social isolation. David does not hate people in general but instead scrupulously respects the rights of everyone, pays his taxes, donates to charity, treats others with politeness, and otherwise fulfills his standard moral duties. Yet it seems that David should not cut himself off from others in this way and should seek out friendships, romantic attachments, community ties, or other loving relationships.

B. Maintaining Loving Relationships

Miguel and Doreen have been happily married for twenty years. Their marriage seems to include moral requirements that are different from or more demanding than what they owe other people in general, such as ones that concern candor, discretion, fidelity, trust, promoting the wellbeing and self-respect of one another, expressing love and respect, and tolerating, apologizing for, and forgiving certain faults or transgressions. Miguel and Doreen also seem to be required in many cases to prioritize one another's interests over their own or those of other people. For example, it seems that Miguel should save Doreen rather than a stranger in a standard lifeboat case; Doreen should defend Miguel from scurrilous and disparaging attacks at some cost to her career; and Miguel should help Doreen to finish her degree rather than donate his time and money to the local food bank. And traditional moral duties that Miguel and Doreen have to people in general

---

[6] Although our ordinary moral judgments about these cases might vary somewhat and depend on further details, the examples exhibit several apparently moral aspects of many loving relationships.

seem to include exceptions that occasionally permit them, for example, to lie, cheat, or steal to get needed medical treatment for each other, to refuse to turn in or testify against one another, and to avoid participating in a just war to care for each other.

C. Perfecting Loving Relationships

Ramari is an active member of a thriving teachers union who feels a deep sense of camaraderie with his fellow members based on their shared commitments and joint projects. He also finds, however, that he envies those who have greater influence than he does; he begrudges certain members for past slights; he sometimes loses his temper; and he scorns those who he thinks are not fully committed to their cause. Ramari, it seems, has moral reasons to improve the bonds of solidarity he has with members of his union by striving, for example, to combat and overcome his corrosive envy, to give up many of his grudges, grievances, and resentments, and to keep his cool more often. He and other members also, it seems, have good moral reasons to institute fair procedures for raising complaints and adjudicating conflicts as well as to employ ceremonies, rituals, and gatherings to express and reinforce their ties with one another.

D. Respecting Loving Relationships

Lisa and June were best friends who recently fell out over a heated series of arguments in which they both said some especially harsh things that they knew their relationship could not survive. Soon after their split, Lisa and June began divulging one another's secrets, such as Lisa's marital problems and struggles with alcohol and June's true political leanings and demotion at work. Lisa and June also tend to focus on and exaggerate the negative aspects of their past friendship, chide themselves for trusting one another for so long, assume that the other person was merely using them, and express these views to others. It seems that Lisa and June should have and show greater respect than this for the thriving friendship they once had by, for example, cherishing and venerating its memory, not regarding their prior relationship as a sham or a waste of time and energy, and not betraying one another's confidences.

E. Promoting Loving Relationships

Finally, Bob sees loving relationships among other people as a weakness he can exploit. He imposes working conditions on his employees that pit them against one another and strain their marriages, fosters envy and enmity among his children, works to alienate his wife from her family, and sows doubt and distrust among his acquaintances. Bob, it seems, should not seek to discourage or prevent those around him from having loving relationships of various kinds but instead should seek to foster such relationships among others.

II. Loving Relationships in Kantian Moral Theory

Can Kantian frameworks appropriately capture and explain these apparent moral dimensions of loving relationships?

Basic features of such frameworks seem to be incompatible with doing so. According to Kantian ways of thinking, all of normativity is ultimately grounded in the nature and operation of the faculty of reason, which determines what things are valuable and worth striving for, what we ought to do, what we have reason to do, and so on. Loving relationships, such as friendships, are not intrinsic values in G. E. Moore's sense; there is not a hodgepodge of independently existing reasons to, for example, promote or maintain them that our faculty of reason merely allows us to recognize and moves us to satisfy; and our natural affections, loyalties, and love for other people do not by themselves make our relationships with them valuable or ground normative requirements or reasons of any kind.[7] Any value, requirements, or other normative aspects of loving relationships, according to Kantian theory, must ultimately arise from reason itself.

That faculty, however, seems to be too formal, thin, and impartial to capture and explain many of the reasons and requirements we seem to have to form, maintain, perfect, respect, and promote loving relationships. Reason is the power we use to do logical proofs and pursue and organize our ends in consistent and efficient ways. On Kantian ways of thinking, reason also includes a fundamental moral standard with its several

---

[7] See Moore (1993) and Scanlon (1998).

formulations that express a universal and impartial concern for all rational people as such and that generate moral requirements about how to treat everyone in general.

Kantians could accept that our frameworks cannot capture and explain many of the apparent reasons and requirements that concern loving relationships. Some Kantians might argue that many of these moral judgments are illusory, point out that it is understandable but unjustified to claim the authority of reason for things we merely naturally care about, claim that our basic philosophical commitments are more firmly grounded than any remaining moral judgments we cannot capture and explain, and emphasize the moral dangers of parochialism and tribalism along with the ideal of a cosmopolitan, universal, and impartial moral standpoint that emphasizes our shared humanity.

How might Kantians nonetheless attempt to capture and explain some of the normative dimensions of loving relationships while still retaining our commitments to the primacy, authority, impartiality, and universality of reason?

A. Duty of Beneficence

One place to look is the duty of beneficence, which Kant characterizes as a wide and imperfect requirement of reason to set the happiness of others as one of our ends.[8] The duty of beneficence does not specify who, when, or how to help, so, as long as we stay within the bounds of our other duties, we are allowed to direct our beneficence to those we love and to prioritize their happiness over that of others.[9] Miguel, for example, can choose to save Doreen rather than a stranger from drowning when both options are permissible because the duty of beneficence allows him to save her on the basis of his natural affections and love for her.[10]

A problem with this approach, however, is that the duty of beneficence only *permits* us to help and to prioritize the happiness of our loved ones, even though it seems that we are sometimes morally *required* to do so.[11] It seems that Miguel, for example, should save Doreen from drowning instead

---

[8] MM 6:388–90, 453–4; Eth-V 27:561–2.   [9] MM 6:451–2.
[10] See Herman (1993a), Baron (2008), Bramer (2010).
[11] For further discussion of the duty of beneficence, including whether it includes some strict requirements to save others from duress, see Hill (2002a), Timmermann (2005), Stohr (2011).

of a stranger in a standard lifeboat case, whereas the duty of beneficence merely permits him to save either person, or perhaps no one at all, as he pleases. A related difficulty is that the duty of beneficence cannot adequately explain why we apparently have special reasons to care about, promote, and prioritize the happiness of our loved ones.[12] When Miguel faces a choice between saving Doreen or a stranger, the mere fact that Doreen is his wife seems to be a reason to save her, whereas the duty of beneficence implies that both actions are merely instances of doing something good for a person and that Miguel is permitted to apportion his beneficence on non-rational grounds, such as his natural desires, natural feelings, and personal ends. And, in Kant's system of moral duties, the duty of beneficence is strictly subordinated to narrow and perfect ones, which we must never violate when promoting the happiness of others, whereas it seems that we are sometimes morally justified in stealing, breaking promises, and lying for the sake of the happiness of our loved ones.

Perhaps the most significant limitation of appealing to the duty of beneficence to capture and explain the various moral dimensions of loving relationships is that, while the duty of beneficence enjoins us to set the happiness of others as an end, many of the apparent moral reasons and requirements to form, maintain, perfect, respect, and promote loving relationships do not, or do not only, concern or reduce to happiness. David's reasons not to socially isolate himself and to seek out loving relationships with others do not seem fully explained by any increase in happiness that others might get from being David's friend. Miguel and Doreen, it seems, should be faithful to one another, keep one another's confidences, defend one another from disparaging treatment, help one another to develop their natural abilities, and show respect to one another, not simply because of the contributions these make to their happiness. And it seems that, apart from considerations of happiness, Ramari should strengthen the bonds of solidarity with members of his union, Lisa and June should show respect for their past friendship, and Bob should not goad others into avoiding or abandoning certain loving relationships.

In light of its wide latitude, its justification in terms of reciprocal concern among all people or the basic rational nature we all share, its subordination to strict moral duties, and its nature as the most basic rational requirement concerning the happiness of everyone in Kant's framework, the duty of

---

[12] Sticker and van Ackeren (2018).

beneficence alone, it seems, cannot capture and explain most of the moral reasons and requirements that loving relationships seem to involve.

B. Instrumentalism

A second strategy for attempting to show that the apparent moral dimensions of loving relationships are part of Kantian moral frameworks is to highlight ways in which such relationships are often effective or essential means for promoting moral ends or complying with moral principles.[13] Kant argues, for example, that establishing, maintaining, and perfecting certain kinds of communities is necessary for human persons to counteract evil influences, enliven our moral dispositions, and otherwise come close to achieving moral perfection.[14] Certain friendships, marriages, associations, and other loving relationships, he also claims, tend to promote enlightenment, peace, morally useful knowledge, general happiness, and the development of our natural abilities.[15]

There are standard concerns, however, with instrumentalist forms of justification, such as that they depend on contingent, potentially changing, and sometimes unknowable causal chains, that they can ground opposing reasons and requirements that outweigh ones that seem to exist, and that they can at best justify only rules of thumb unless some attitudes or actions are necessary means for promoting some end. For example, there might be better ways to promote enlightenment, peace, and natural self-perfection than being in loving relationships with others; relationships often engender envy, rivalry, parochialism, and other attitudes and actions that undermine rather than promote various morally good ends; and violating the requirements of relationships is sometimes the most effective way to further such ends.

Even if an instrumentalist approach allows Kantian moral theories to capture most or all of the reasons and requirements that loving relationships seem to involve, a remaining concern is that these explanations do not fully accord with *why* we seem to have these reasons and seem to be subject to these requirements. For example, it seems that David should form loving

---

[13] Stark (1997), Cureton and Hill (2018). A stronger claim that some Kantians defend is that reasoning, rational agency, and standards of rationality are constitutively social in the sense that they presuppose other people who we can give reasons and arguments to, seek their rational consent, and so on (O'Neill 1989; Korsgaard 1996c, lecture 4; Herman 2007, ch. 1).

[14] Rel 6:96–8.   [15] MM 6:473; Anth 7:277; WOT 8:144.

relationships with others, that Miguel and Doreen owe it to one another to be candid, discreet, and faithful, and that Lisa and June should not disparage their prior friendship whether or not these actions happen to improve their knowledge, help them to develop their natural abilities, or otherwise promote morally good ends. If spilling our friend's deepest secrets, cheating on our spouse, cutting ourselves off from others, and demeaning a past friend were the only ways to promote various moral ends, it seems that sometimes we should nonetheless refrain from doing these things.[16]

These doubts about whether the duty of beneficence and instrumentalist forms of justification, together or separately, allow Kantian theories to capture and explain the apparent moral aspects of loving relationships might be overcome, but they suggest that a more radical and unorthodox approach is likely needed for Kantians to square our commitment to the primacy of impartial reason with the morally appropriate partiality that such relationships seem to involve.

III. Reason and Its Interests

Throughout his published and unpublished writings as well as in lecture notes from his courses, Kant explores and develops an underappreciated and potentially fruitful theme concerning the nature of reason, which is that rational nature in each of us has its own interests that help to determine what moral principles it legislates. Although attributing this theme to Kant is likely to meet with significant skepticism from many Kantians, it opens new possibilities for interpreting, reconstructing, or supplementing Kant's moral framework in ways that can capture and explain many of the reasons and requirements that some loving relationships seem to involve.

A rational interest or interest of reason is a goal, aim, or need that the faculty of reason, on its own, endorses and moves us to satisfy. Reason, according to Kant, is an active mental faculty with its own needs, objects, and ends that differ from and sometimes conflict with one another and with our natural desires and chosen ends. The things that reason takes an interest in acquire their normative status by the nature and operation of reason itself.

---

[16] By analogy, if murdering someone were necessary to promote a moral value, such as general welfare, perpetual peace, or even an ideally moral world, then in at least some cases we should not do so, although this might depend in some ways on how much good we can produce by putatively immoral actions.

This basic feature of reason accords with ordinary ideas of a fully rational and reasonable person who, as such, seems to care about certain things apart from what she might otherwise happen to want or choose.

One rational interest, according to Kant, is in doing our duty from duty, which is an aim of the faculty of reason and something it moves us to do. The faculty of reason, Kant sometimes suggests, also has interests in explaining and systematically unifying things, making them consistent and harmonious, preserving, protecting, developing, and exercising rational nature, protecting freedom, promoting happiness, acquiring knowledge, communicating with other people, respecting them and showing them respect, self-development, equality, autonomy, and, as I will suggest in the next section, a kind of solidarity among people.[17] These are innate, intrinsic, and potentially conflicting interests that each of us, Kant claims, is moved by our faculty of reason to satisfy for their own sake apart from whatever natural desires and feelings we might also have.

In addition to providing a novel way to interpret some of Kant's ideas and to address longstanding philosophical questions regarding the nature of reason and what it is to be a rational or reasonable person, ascribing these and perhaps other interests to the faculty of reason itself also suggests a possible structure for Kantian moral theories that allows them to capture and explain a wider variety of reasons and requirements than they otherwise could. Assuming that the Categorical Imperative is interpreted as a principle of justifiability to all, these rational interests provide standards and criteria for whether each of us could rationally will a maxim as a universal law or whether everyone would rationally legislate a candidate principle for an ideal Kingdom of Ends. In addition to standards of rational prudence that Onora O'Neill, Christine Korsgaard, and Barbara Herman regard as standards of rational willing, my radical suggestion is that these standards also include our many formal and substantive rational interests.[18]

The theme I have partially sketched needs further elaboration, defense, and textual support. One way to develop it is by focusing on specific interests of reason that Kant describes. Let's consider, in particular, a rational interest in a kind of solidarity and examine whether that interest, combined with a

---

[17] e.g., A307/B364, A305/B361, CPJ 5:294, Rel 6:58, WOT 8:146n, Anth 7:265; MM 6:237, CPrR 5:61; G 4:430, A644/B672, MM 6:471–2, CPrR 5:77, G 4:423, MM 6:237–8, WOT 8:145, and Eth-C 27:429, respectively.

[18] O'Neill (1989, 91–3), Korsgaard (1996b), Herman (1993b, 121–2).

principle of justifiability to all, can ground reasons and requirements to form, maintain, perfect, respect, and promote relationships of that kind.

IV. Solidarity

We can begin to explore rational interests in solidarity by first considering the nature of solidarity. Kant's discussions of loving relationships, such as friendship, marriage, family, and community, are scattered, often incomplete, and sometimes apparently contradictory, but we can draw on and abstract from these discussions to characterize a general type of solidarity that includes some relationships of those other kinds.[19] This conception of solidarity has four paradigmatic features.

A. Shared Rational Commitments

The first paradigmatic feature of solidarity is that a group of people each share an effective commitment that is favored by reason. Commitments are stable choices that include ends, maxims, policies, plans, and projects. Someone has an effective commitment only if she tends to live up to it. Two or more people share a commitment if they are committed to the same thing, such as a common goal or principle. And the faculty of reason favors commitments if they serve one or more of our rational interests, such as interests in consistency, harmony, communication, happiness, acquiring knowledge, and affirming and acting from moral principles.

Miguel and Doreen, for example, are committed to promoting one another's happiness.[20] The faculty of reason favors this shared commitment because of its intrinsic interests in the happiness of all. Ramari's teachers union is a group of people who endorse many of the same genuine moral principles and goals, who aim for all of them to live up to those commitments, and who work together to eliminate obstacles, to help one another develop strength of will, and to excite "the moral incentives of each individual."[21] Before their split, Lisa and June shared an effective commitment to

---

[19] For further discussion of Kant's accounts of various kinds of relationships, see Herman (1993a), Paton (1993), Korsgaard (1996a, 215–16), (Wood 1999, ch. 8), Denis (2001), Ebels-Duggan (2009), Guyer (2011).
[20] Eth-C 27:425.   [21] Rel 6:197; cf. Rel 6:93, 95, 124, 151; MM 6:469; Eth-V 27:677, 682.

open communication with one another that the faculty of reason favors because of its intrinsic interest in communication as well as its derivative interest in communication as a means of promoting knowledge and correcting errors.[22] David could join a community orchestra that is committed to developing the musical abilities of everyone in their group.[23] Bob might be preventing his employees from forming relationships based on their shared commitments to advancing and promulgating the scientific research that their company produces.[24] The faculty of reason favors these commitments because of its intrinsic interests in natural perfection and in promoting knowledge.[25] And, more generally, the faculty of reason, in at least one respect, favors shared commitments as such because of its formal interests in unity, which includes interests in promoting and maintaining convergence among the commitments of different people, whatever those commitments happen to be.[26]

B. Trust in Shared Rational Commitments

The second paradigmatic feature of solidarity is that each of the people trusts that they all share an effective commitment that reason favors.[27] To trust that someone has a commitment of this sort, according to Kant, is to judge with conviction that she endorses and will likely maintain and live up to an end, aim, project, or principle that serves an interest of reason. Trust of this kind can be reasonable in two ways.

First, we might know that someone endorses a commitment if our judgment is based on mental states that represent grounds that indicate that the judgment is probably true, such as credible testimony from others or first-hand experiences. And, second, we might reasonably hope that someone has a commitment on the basis of non-representational mental states that arise from our faculty of reason itself, such as ones that might lead us to judge with conviction that, without good evidence to the contrary, other people are honest and good.[28]

Developing reasonable trust among a group of people often requires them to show one another that they endorse the same commitments. An orchestra

---

[22] Eth-V 27:683.   [23] Eth-C 27:428; Eth-V 27:679.   [24] L-Anth 25: 702, 1347.

[25] Kant emphasizes that we have no *duty* to promote the natural perfection of others, but this is compatible with reason nonetheless taking an interest in the natural perfection of all.

[26] Eth-V 27:681–3, 703; Eth-H 27:50; Eth-C 27:429.   [27] Eth-V 27:681; Eth-C 27:429.

[28] L-Log 24:246.

might express their shared commitment to developing their natural abilities by practicing together for long hours, choosing complicated pieces, mentoring younger members, and visibly taking pleasure in the group's achievements. Lisa and June showed their commitment to open communication between them by confiding in one another. And the members of the teachers union developed a "moral bond" with one another by, in part, each ensuring "that his actions not only furnish a negative example, in containing nothing evil, but also provide a positive one, in possessing an element of good."[29]

Social structures, such as norms, rules, laws, formalities, ceremonies, observances, and traditions, can also provide ways for people to develop trust in one another. For example, Ramari's teachers union regularly holds public assemblies in which their common cause is "loudly proclaimed and thereby fully shared"; they maintain "this fellowship through repeated public formalities which stabilize the union of its members;" and they use ceremonies and instruction as ways of "transmitting" their shared commitments and trust "to posterity through the reception of new members."[30] Their leaders regularly make speeches "in the name of the whole" group in order to make its shared concerns "visible as a public issue" so that the wishes of each person in the group are "represented as united with the wishes of all toward one and the same end."[31] And they establish and enforce rules that they trust one another to follow.[32] Miguel and Doreen also use rules, ceremonies, and other social structures to maintain and enhance their trust in one another by, for example, regularly cooking together, calling ahead when one of them will be late, celebrating holidays and anniversaries, and spending Sunday afternoons together.

Developing and maintaining trust that someone else shares our commitments can nonetheless be difficult because we might not be sure whether she is sufficiently self-aware to know her own commitments, whether she has the commitment or is merely pretending to have it, and whether she endorses it on rational grounds. David, we can imagine, came to lose trust in his ex-wife when he discovered that they affirm "quite different principles" and realized that he is "utterly opposed to" some of her basic values and commitments.[33]

---

[29] Eth-C 27:412.   [30] Rel 6:193.   [31] Rel 6:197.   [32] MM 6:307.
[33] Eth-V 27:682.

C. Love

A third paradigmatic feature of solidarity is that each of the people is committed to the happiness of the others for their own sakes because of their reasonable and mutual trust that they all share an effective commitment that reason favors.[34] A commitment to the happiness of others for their own sake includes adopting it as a non-instrumental end, tending to prioritize that end over the happiness of other people, and regularly choosing to act in ways that promote it. A group of people might be committed to one another's happiness on other grounds, but this special practical love for one another arises from and is sustained by their reasonable trust in the shared rational commitments of one another.[35] Union members, for example, might have a love for one another for their own sake that is derived from their trust in one another to support the cause, but if this trust disappears then they lose this commitment to the happiness of their comrades while still perhaps maintaining their general love for other people as such.

An effective and easily recognizable way to implement our commitment to the happiness of someone we are in solidarity with is to advance, promote, or live up to the shared goals, principles, projects, or other commitments that provide the bases for our mutual love. Each of us in a relationship of this sort began with a commitment that reason favors, but the love for one another that arose on the basis of our trust that we share this commitment can provide an additional ground for endorsing and maintaining that commitment as a way of promoting the happiness of our comrades. These additional grounds might increase the priority we give to the commitment as compared to others we endorse and lead us to resist, for example, undermining or violating the commitment because doing so would hurt and betray those we care about. How much priority we give to promoting the happiness of a comrade over that of other people or even over our own happiness can also vary "because the limits here are not defined, and there can be no indication of degree as to how far I ought to care for myself, and how far for others," so that the measure of solidary dispositions "is not determinable by any law or rule."[36]

---

[34] MM 6:469, 471; Eth-V 27:676–7, 682–4; Eth-C 27:424–5.
[35] MM 6:452; Eth-C 27:424.   [36] Eth-C 27:424.

D. Trust in Love

And a fourth paradigmatic feature of solidarity is that each of the people reasonably trusts that they all share a commitment to one another's happiness for its own sake on the basis of their mutual trust that they all share an effective commitment that reason favors.[37] It is "in itself reassuring to be able to count on . . . assistance" from a comrade, to "confidently count on the other's help . . . in case of need," and to "have confidence . . . that he would be able and willing to look after my affairs."[38]

Developing this kind of public trust among a group of people often requires them to show one another that they are committed to one another's happiness. Miguel and Doreen might, for example, assure one another that they stand ready to help in times of need, show concern for one another's misfortunes, do small favors for one another, give gifts, and promote their common goals. The teachers union uses social structures of various kinds as ways of developing and securing trust in their love for one another, such as rules about supporting members who are on strike or in hospital, celebrating birthdays, and "a ritual communal partaking at the same table" that represents a kind of "brotherly love" among them.[39]

There are various difficulties with creating and sustaining this public trust that arise from fears that others are merely using us or even secretly hate us. When they were friends, for example, June regularly demanded Lisa's help and burdened her with her troubles, which led Lisa to worry that June is "ungenerous" towards her and merely out for herself, does not love her for her own sake, and only aims in their relationship "to secure some attention to [her own] needs."[40] Lisa even began to wonder whether June hates her. On some occasions, "in a fit of anger," June would "consign [Lisa] to the gallows", heap "coarse rebukes" on her, but also offer "apologies the moment [she] calms down."[41] June also showed signs of envy for Lisa's accomplishments and merits that suggested that her aim was to bring Lisa down rather than pull herself up. And June sometimes used Lisa's confidences against her.[42] Miguel and Doreen, on the other hand, have an expressed willingness to forgo help from the other person on some occasions, not to "cause

---

[37] Eth-V 27:676; Eth-C 27:426.
[38] Eth-V 27:684, MM 6:471, and Eth-C 27:425, respectively.   [39] Rel 6:199.
[40] Eth-C 27:425; cf. MM 6:471 and Eth-C 27:425, respectively.
[41] Eth-C 27:430; Eth-V 27:685.   [42] Eth-C 27:427.

trouble" to one another, to "endure [certain of their] woes alone", and "not make demands" on one another's help, as ways of showing their love for each other.[43] Solidarity is one form of special tie in which we basically share rational commitments with one another, trust that we share these commitments, love one another on the basis of this trust, and trust that we love one another. There is an ideal form of solidarity in which the four paradigmatic features are fully satisfied but also imperfect ones in which these features are satisfied to varying degrees. Various kinds of more specific relationships can be forms of solidarity, such as friendships, marriages, communities, and so on, although these relationships have other features that differentiate them from one another.

V. Rational Interests in Solidarity

The faculty of reason, Kant sometimes suggests or implies, has intrinsic interests in forming, maintaining, perfecting, respecting, and promoting relationships of solidarity. He describes some relationships of this sort as "practically necessary" ideas and ideals of reason that make us "deserving of happiness," that serve "a purely intellectual need" of reason itself, that further interests of "humanity," and that "should inspire respect."[44] These passages, however, concern various competing and fragmentary conceptions of friendship. Kant also valorizes other kinds of relationships, such as marriages, families, and communities.[45] And he says more generally that the "human being is destined by his reason to live in a society with human beings" and that humans have "a calling to use their reason socially."[46]

Aside from this explicit textual evidence, which is inconclusive, some of the specific duties that Kant describes seem to presuppose that the faculty of reason has interests in relationships of solidarity. We will consider some of these duties in the next section, but from a broadly Kantian perspective and a common-sense standpoint, it seems that a rational and reasonable person would, as such, favor relationships in which people share rational commitments, trust that they have these commitments, love one another on this basis, and trust that they have this mutual love. Relationships of solidarity

---

[43] Eth-C 27:425, cf. Eth-V 27:684; MM 6:471.
[44] MM 6:469, Eth-V 27:680–2, and Eth-C 27:429, respectively.
[45] Eth-V 27:493; Rel 6:193; Ped 9:494.   [46] Anth 7:325 and L-Log 24:151, respectively.

incorporate and arrange a variety of other interests that reason has, such as in committing to ends, principles, or projects that it favors, in committing to promote the happiness of other people, in expressing these rational commitments, and in trusting other people. And attributing this interest to the faculty of reason allows us to capture and explain many of the reasons and requirements that such relationships seem to involve.

Aside from principles of logic, it is notoriously difficult to explain why something is a feature of reason. Kant's strict, a priori, and transcendental methods for doing so, which he mainly develops in the *Critiques*, might not justify some of the rational interests that he himself describes. Perhaps the best we can do is develop candidate conceptions of reason, draw out their normative implications, and eventually hope to assess them as a whole by how well they capture and explain our considered judgment concerning, for example, the nature of a rational and reasonable person, what kinds of reasons we have, and ordinary ways of speaking, thinking about, and appealing to reason.

VI. *Prima facie* Laws of Reason

Let's suppose that the faculty of reason has intrinsic interests in each of us establishing relationships of solidarity with other people, maintaining and perfecting ones we are in, promoting such relationships among other people, and respecting relationships of solidarity themselves. What sorts of *prima facie* laws or principles might these interests favor when combined with a principle of justifiability, which requires us to conform to laws that are justifiable to everyone on the basis of our interests of reason? The principles I mention below, which in most cases Kant himself endorses in some form and on some occasions, are *prima facie* principles that reason favors in the sense that, all else being equal, each of us could or would rationally will them, or ones like them, as universal laws on the basis of our rational interests in solidarity.

A. Prima Facie *Laws about Forming Relationships*

Our rational interest in forming relationships of solidarity leads reason in each of us to favor various kinds of presumptive or *prima facie* laws, including "a duty to oneself as well as to others not to isolate oneself" from all other people.[47] It also favors a prohibition on hating and "shying away from human beings" in general in ways that make it difficult or impossible for us to develop relationships of solidarity with them.[48] These laws also forbid more limited forms of misanthropy in which we hate or isolate ourselves from anyone outside of our family, tribe, or community, such as a sect that attempts to "cut itself off from all other peoples and avoid intermingling with them," because doing so makes it difficult or impossible for us to form relationships of solidarity with those other people.[49]

Misanthropy tends to arise, according to Kant, by overgeneralizing from particular cases, such as people like David who have been "cheated," have been "ill-used for their benevolence," or have observed other forms of vice and immorality and so come to "trust no other human being."[50] In light of this tendency, our rational interest in forming relationships of solidarity also favors indirect laws that combat misanthropy, such as ones that forbid us from overgeneralizing in these ways and that require us to diminish instances that are especially likely to engender misanthropy in others, such as "[f]alsehood, ingratitude, injustice."[51]

These presumptive laws also require us to strive for solidarity with other people by, for example, adopting commitments that reason approves of, searching for other people who share those commitments, and communicating with them in ways that promote mutual trust and reciprocal love. They require us to develop and express traits that tend to lead others to form relationships of solidarity with us, such as "uprightness of disposition, candor and trustworthiness", "conduct that is free from malice and falsity," and "vivacity, amiability and cheerfulness of mind."[52] There might be limits, however, to how many relationships of solidarity we can form with others.[53]

And these presumptive laws require us to treat everyone, including our enemies, as if they might someday stand in a relationship of solidarity with us. Acting from this maxim is "a course of conduct appropriate to the use of reason, and conformable to the laws of morality" because all human beings are worthy of being in these relationships with us, treating them in this way tends to make it more likely that we will enter into such relationships with them, and, if we manage to form such a relationship with someone, "we do a

---

[47] MM 6:473; cf. MM 6:402; MM 6:471.
[48] MM 6:466; cf. TP 8:307; CPJ 5:276; Rel 6:34.   [49] Rel 6:184.
[50] Anth 7:205, Eth-C 27:440, CPJ 5:276; MM 6:466; L-Anth 25:553, and Anth 7:205, respectively.
[51] CPJ 5:276.   [52] Eth-C 27:429.   [53] Eth-V 27:673, 685.

service to mankind, or to humanity" by establishing a relationship that reason approves of. More specifically, we should not, as Lisa and June did, disparage people with whom we were previously in a relationship of solidarity to third parties, not only because we might someday reconcile with our prior comrades but also because this tends to make "those to whom we say such things" avoid forming relationships of solidarity with us out of fear that "the same might happen to them" if we form a solidary relationship with one another but later fall out.[54]

B. Prima Facie *Laws about Maintaining Relationships*

The rational interest in maintaining and not undermining relationships of solidarity we are in favors laws of various kinds. Relationships of solidarity, according to Kant, ground special duties that the parties have to one another because their shared goals, projects, principles, or other commitments, along with their commitments to the happiness of one another, are "common and simultaneous."[55]

One kind of duty they have is to maintain and live up to the commitments they share by, for example, promoting and not undermining their common goals, complying with their shared principles, and pursuing their joint projects. For example, if Miguel and Doreen are in a solidary relationship with regard to their shared aim of having and raising a child "as their joint work," then, if they produce a child together, they incur "an obligation . . . towards each other to maintain it."[56] Or "pure sincerity in friendship can be no less required of everyone even if up to now there may never have been a sincere friend, because this duty—as duty in general—lies, prior to all experience, in the idea of a reason determining the will by means of a priori grounds."[57] Friends also have a duty to be candid with one another as a way of living up to their shared commitment to open communication.[58]

A second kind of duty concerns promoting the happiness of our comrades. When people are in relationships of solidarity, there are "certainly duties to which they are obligated," such as to help one another in times of need, not to wish for or take pleasure in the misfortune of one another, not to "misuse [the] trust" that they have in one another's goodwill, not to allow the other person to help us without being "generous in [our] turn," and

---

[54] Eth-V 27:680–1.   [55] Eth-V 27:696.   [56] MM 6:381.   [57] G 4:408.
[58] Rel 6:33.

"participating and sharing sympathetically in the other's well-being."[59] In some cases, for example, it is "a duty for one of the friends to point out the other's faults to him; this is in the other's best interests and is therefore a duty of love," but in other cases to "uncover his weaknesses" or to "censure his errors" is "contrary to the duty of friendship" because doing so would "injure his self-love."[60] Among friends who are committed to complete candor with one another, each of them is "bound not to share the secrets entrusted to him with anyone else, no matter how reliable he thinks him, without explicit permission to do so."[61] A brother might be under a presumptive law of "kinship" not to serve as a witness against his sibling in ways that would harm him even though he might be under other presumptive laws of greater priority to be candid with the authorities.[62] And if a "wife loses her husband, then the grown-up, well-behaved son has the duty incumbent on him, and also the natural inclination within him, to honor her, to support her, and to make her life as a widow pleasant."[63]

Maintaining relationships of solidarity also involves not undermining the four paradigmatic features of those relationships, so we "must not", for example, "seek to diminish" the "well-wishing dispositions" the other person has towards us or the trust the other has in our goodwill towards them by, for example, leading them to think that we do not love them or that we are simply using them to fulfill our selfish interests.[64] Indiscreetly sharing a friend's secret, for example, likely diminishes her confidence that we love her, while violating the moral standards of our community tends to diminish the trust others may have that we share their moral commitments.

And our rational interest in maintaining relationships of solidarity favors laws that require us on some occasions to apologize to our comrades, reaffirm our commitments and love for them, and refrain from quarreling in ways that make "mutual trust impossible during a future peace."[65] On some occasions, however, "one must break off the association that existed or avoid it as much as possible."[66]

---

[59] Eth-V 27:696, Rel 6:33, Eth-C 27:426, and MM 6:471, respectively.
[60] MM 6:470 and Eth-V 27:685, respectively.   [61] MM 6:472. See Flynn (2007).
[62] Eth-V 27:493.   [63] Anth 7:310.   [64] Eth-C 27:426.
[65] TPP 8:346; cf. MM 6:471. This quote refers to actions of states during times of war, but many of the Articles of Perpetual Peace seem to have plausible analogues in interpersonal relationships.
[66] MM 6:474, cf. MM 6:365; Anth 7:294; Eth-C 27:425; L-Anth 25:1390.

C. Prima Facie *Laws about Perfecting Relationships*

Our rational interest in perfecting our solidary relationships favors laws that require us to adopt and strive to realize ideal solidarity.[67] This ideal consists in each of the people sharing exactly the same commitment that reason maximally approves and affording it the same high priority, fully trusting that this is the case, maximally loving each other, and fully trusting that this is so.

In most relationships of solidarity, however, one or more of these elements is not fully realized. We often do not know for sure what exact commitments and attitudes we or others have or what priority we or they assign to them.[68] Perfecting our relationships of solidarity thus often requires us to "track down . . . any misunderstandings that hinder agreement; to clear up errors and come together as much as possible" with our comrades, as well as to communicate the degree of love we have for one another.[69] Perfecting our relationships of solidarity might also involve committing more fully to ends, goals, or principles that we share with others as well as to promoting their happiness and striving to eliminate feelings of anger, resentment, or envy towards our comrades that tend to undermine these commitments.[70]

D. Prima Facie *Laws about Respecting Relationships*

Our rational interest in respecting relationships of solidarity leads it to favor laws that require us to "venerate" relationships of this kind, which "should inspire respect" in us.[71] Respecting relationships of solidarity involves judging that they are good in themselves and not judging that they are merely useful or contemptible.[72] We are also not supposed to act in ways that express lack of respect or disrespect for such relationships. For example, it is "bad in itself to speak disparagingly" of a previous friend, even if he has become our enemy, "in that we thereby show that we have no respect" for our relationship.[73] And we must not show disrespect for "even the memory

---

[67] MM 6:469, 471.   [68] MM 6:471.   [69] Eth-V 27:685.
[70] MM 6:471; Eth-V 27:678–9; Ped 9:484–5.   [71] Eth-C 27:429.   [72] MM 6:479.
[73] Eth-C 27:429.

of a friendship now broken off" by, for example, "abusing later on the former confidence and candor of the other person."[74]

E. Prima Facie *Laws about Promoting Relationships*

Finally, our rational interest in promoting relationships of solidarity among other people leads it to favor laws that require us, for example, not to act in ways that prevent or undermine such relationships. Teachers, for example, "must not prefer one child over another because of its talents but only because of its character, for otherwise resentment develops, which is contrary to friendship." It is also "wrong" for teachers to oppose friendships among children because "[t]he child must maintain friendships with others and not remain by itself all the time."[75] We should not tempt people to violate the duties of their relationships, prevent them from associating with one another, or denigrate their relationships of solidarity. And we should provide opportunities for them to form relationships of solidarity, encourage them to develop traits that help them to do so, and do what we can to promote mutual trust and love among them.[76]

VII. Relationships of Solidarity in an Ideally Rational World

A longstanding problem for Kantian theory is to reconcile impartial reason with the partiality the moral life seems to involve. I have suggested that the faculty of reason itself has interests in a kind of solidarity that, when combined with an impartial principle of justifiability, generates in a plausible way many of the presumptive requirements and reasons that such relationships seem to involve, including those in the five examples I began with. A fully developed theory of reason along these lines, however, must go on to specify our other rational interests and their relative priorities as well as interpret the principle of justifiability in ways that allow us to adjudicate conflicts among the rational interests of different people.

In a cryptic passage from lecture notes on Kant's winter 1784–5 course in moral philosophy, Collins reports that Kant said:

---

[74] Anth 27:194.   [75] Ped 9:484–5.   [76] Ped 9:499.

Friendships are not found in heaven, for heaven is the ultimate in moral perfection, and that is universal; friendship, however, is a special bond between particular persons; in this world only, therefore, it is a recourse for opening one's mind to the other and communing with him, in that here there is a lack of trust among men.[77]

If, as Kant sometimes suggests, we have rational interests in solidarity, then Kant could have drawn a different conclusion, namely that a fully rational world would include many friendships, families, associations, communities, and other relationships of solidarity as well as itself constitute a "unity of humankind as that of a family" in which there is public knowledge among good people that they love and trust one another on the basis of their shared commitments to the laws and principles of reason itself.[78]

Works Cited

Annas, Julia. 1984. "Personal Love and Kantian Ethics in Effi Briest." *Philosophy and Literature* 8 (1): 15–31.

Baron, Marcia. 2008. "Virtue Ethics, Kantian Ethics, and the 'One Thought Too Many' Objection." In *Kant's Ethics of Virtues*, ed. Monika Betzler, 245–78. Berlin: De Gruyter.

Bramer, Marilea. 2010. "The Importance of Personal Relationships in Kantian Moral Theory: A Reply to Care Ethics." *Hypatia* 25 (1): 121–39.

Cureton, Adam, and Thomas E. Hill, Jr. 2018. "Kant on Virtue: Seeking the Ideal in Human Conditions." In *The Oxford Handbook of Virtue*, ed. Nancy Snow, 263–280. Oxford: Oxford University Press.

Denis, Lara. 2001. "From Friendship to Marriage: Revising Kant." *Philosophy and Phenomenological Research* 63 (1): 1–28.

Ebels-Duggan, Kyla. 2009. "Moral Community: Escaping the Ethical State of Nature." *Philosophers' Imprint* 9.

Ferrarin, Alfredo. 2015. *The Powers of Pure Reason: Kant and the Idea of Cosmic Philosophy*. Chicago: University of Chicago Press.

Flynn, Patricia C. 2007. "Honesty and Intimacy in Kant's Duty of Friendship." *International Philosophical Quarterly* 47 (4): 417–24.

---

[77] Eth-C 27:428.   [78] Ped 9:494.

Guyer, Paul. 2011. "Kantian Communities: The Realm of Ends, the Ethical Community, and the Highest Good." In *Kant and the Concept of Community*, ed. Charlton Payne and Lucas Thorpe, 88–120. Woodbridge: Boydell & Brewer.

Held, Virginia. 2006. *The Ethics of Care: Personal, Political, and Global*. Oxford: Oxford University Press.

Herman, Barbara. 1993a. "Could It Be Worth Thinking About Kant on Sex and Marriage?" In *A Mind of One's Own: Feminist Essays on Reason and Objectivity*, ed. Louise Antony and Charlotte Witt, 49–68. New York: Westview Press.

Herman, Barbara. 1993b. *The Practice of Moral Judgment*. Cambridge, MA: Harvard University Press.

Herman, Barbara. 2007. "A Cosmopolitan Kingdom of Ends." In *Moral Literacy*, 51–78. Cambridge, MA: Harvard University Press.

Hill, Thomas E., Jr. 2000. "Donogan's Kant." In *Respect, Pluralism and Justice: Kantian Perspectives*, 119–152. Oxford: Oxford University Press.

Hill, Thomas E., Jr. 2002a. "Meeting Needs and Doing Favors." In *Human Welfare and Moral Worth: Kantian Perspectives*, 201–43. Oxford: Oxford University Press.

Hill, Thomas E., Jr. 2002b. "Reasonable Self-Interest." In *Human Welfare and Moral Worth: Kantian Perspectives*, 125–63. Oxford: Oxford University Press.

Kant, Immanuel. 1992. *Lectures on Logic* [L-Log], ed. and trans. J. Michael Young. Cambridge: Cambridge University Press.

Kant, Immanuel. 1996a. "Groundwork of the Metaphysics of Morals" [G]. In *Practical Philosophy*, ed. and trans. Mary J. Gregor, 37–108. Cambridge: Cambridge University Press.

Kant, Immanuel. 1996b. "The Metaphysics of Morals" [MM]. In *Practical Philosophy*, ed. and trans. Mary J. Gregor, 353–603. Cambridge: Cambridge University Press.

Kant, Immanuel. 1998. *Critique of Pure Reason* [A,B], ed. and trans. Paul Guyer and Allen W. Wood. Cambridge: Cambridge University Press.

Kant, Immanuel. 1999a. "On the Common Saying: That May Be Correct in Theory, but It Is of No Use in Practice" [TP]. In *Practical Philosophy*, ed. and trans. Mary J. Gregor, 273–310. Cambridge: Cambridge University Press.

Kant, Immanuel. 1999b. "Toward Perpetual Peace" [TPP]. In *Practical Philosophy*, 311–52. Cambridge: Cambridge University Press.

Kant, Immanuel. 2000. *Critique of the Power of Judgment* [CPJ], trans. Paul Guyer and Eric Matthews, ed. Paul Guyer. Cambridge: Cambridge University Press.

Kant, Immanuel. 2001a. "Kant on the Metaphysics of Morals: Vigilantius's Lecture Notes" [Eth-V]. In *Lectures on Ethics*, ed. Peter Heath and J. B. Schneewind, 249–452. Cambridge: Cambridge University Press.

Kant, Immanuel. 2001b. "Kant's Practical Philosophy: Herder's Lecture Notes" [Eth-H]. In *Lectures on Ethics*, ed. Peter Lauchlan Heath and J. B. Schneewind, 1–36. Cambridge: Cambridge University Press.

Kant, Immanuel. 2001c. "Moral Philosophy: Collins's Lecture Notes" [Eth-C]. In *Lectures on Ethics*, ed. Peter Heath and J. B. Schneewind, 37–222. Cambridge: Cambridge University Press.

Kant, Immanuel. 2001d. "Religion within the Boundaries of Mere Reason" [Rel]. In *Religion and Rational Theology*, ed. Allen W. Wood and George di Giovanni, 39–216. Cambridge: Cambridge University Press.

Kant, Immanuel. 2001e. "What Does It Mean to Orient Oneself in Thinking?" [WOT]. In *Religion and Rational Theology*, ed. Allen W. Wood and George Di Giovanni, 1–14. Cambridge: Cambridge University Press.

Kant, Immanuel. 2005. *Notes and Fragments* [NF], trans. Curtis Bowman, Paul Guyer, and Frederick Rauscher, ed. Paul Guyer. Cambridge: Cambridge University Press.

Kant, Immanuel. 2007a. "Anthropology from a Pragmatic Point of View" [Anth]. In *Anthropology, History, and Education*, ed. Günter Zöller and Robert B. Louden, 227–429. Cambridge: Cambridge University Press.

Kant, Immanuel. 2007b. *Critique of Practical Reason* [CPrR], ed. and trans. Mary J. Gregor. Cambridge: Cambridge University Press.

Kant, Immanuel. 2007c. "Lectures on Pedagogy" [Ped]. In *Anthropology, History, and Education*, ed. Günter Zöller and Robert B. Louden, 486–527. Cambridge: Cambridge University Press.

Kant, Immanuel. 2013. *Lectures on Anthropology* [L-Anth], trans. Robert R. Clewis, Robert B. Louden, G. Felicitas Munzel, and Allen W. Wood, ed. Allen W. Wood and Robert B. Louden. Cambridge: Cambridge University Press.

Kleingeld, Pauline. 1998. "The Conative Character of Reason in Kant's Philosophy." *Journal of the History of Philosophy* 36 (1): 77–97.

Korsgaard, Christine M. 1996a. *Creating the Kingdom of Ends*. Cambridge: Cambridge University Press.

Korsgaard, Christine M. 1996b. "Kant's Formula of Universal Law." In *Creating the Kingdom of Ends*, 77–105. Cambridge: Cambridge University Press.

Korsgaard, Christine M. 1996c. *The Sources of Normativity*. Cambridge: Cambridge University Press.

McDowell, John. 1994. *Mind and World*. Cambridge, MA: Harvard University Press.

Moore, G. E. 1993. *Principia Ethica*. Ed. Thomas Baldwin. Rev. ed. Cambridge: Cambridge University Press.

O'Neill, Onora. 1989. *Constructions of Reason*. Cambridge: Cambridge University Press.

Paton, H. J. 1993. "Kant on Friendship." In *Friendship: A Philosophical Reader*, ed. Neera Badhwar, 133–54. Ithica, NY: Cornell University Press.

Raedler, Sebastian. 2015. *Kant and the Interests of Reason*. Berlin: De Gruyter.

Rawls, John. 1999. "Kantian Constructivism in Moral Theory." In *Collected Papers*, ed. Samuel Freeman, 303–59. Cambridge, MA: Harvard University Press.

Scanlon, T. M. 1998. *What We Owe to Each Other*. Cambridge, MA: Belknap Press of Harvard University Press.

Stark, Cynthia A. 1997. "Decision Procedures, Standards of Rightness and Impartiality." *Noûs* 31 (4): 478–95.

Sticker, Martin, and Marcel van Ackeren. 2018. "The Demandingness of Beneficence and Kant's System of Duties." *Social Theory and Practice* 44 (3): 405–36.

Stohr, Karen. 2011. "Kantian Beneficence and the Problem of Obligatory Aid." *Journal of Moral Philosophy* 8 (1): 45–67.

Timmermann, Jens. 2005. "Good but Not Required?—Assessing the Demands of Kantian Ethics." *Journal of Moral Philosophy* 2 (1): 9–27.

Velkley, Richard. 2014. *Freedom and the End of Reason: On the Moral Foundation of Kant's Critical Philosophy*. Chicago, IL: University of Chicago Press.

Williams, Bernard. 1981. "Persons, Character and Morality." In *Moral Luck: Philosophical Papers 1973–1980*, 1–19. Cambridge: Cambridge University Press.

Wolf, Susan. 1992. "Morality and Partiality." *Philosophical Perspectives* 6: 243–59.

Wolf, Susan. 2012. "One Thought Too Many: Love, Morality, and the Ordering of Commitment." In *Luck, Value, and Commitment: Themes from the Ethics of*

Bernard Williams, ed. Ulrike Heuer and Gerald Lang, 71–94. Oxford: Oxford University Press.

Wood, Allen W. 1999. *Kant's Ethical Thought.* Cambridge: Cambridge University Press.

Yovel, Yirmiyahu. 1989. "The Interests of Reason: From Metaphysics to Moral History." Seventh Jerusalem Philosophy Encounter. Dordrecht: Springer.

4

The Individualist Objection

or Why Ex Ante Probabilities Aren't Always Individualistic

Jessica J. T. Fischer

Should we be able to access benefits, based on our membership in a group? At first blush, such benefits may strike us as uncontroversial. One may plausibly appeal, for instance, to one's membership in the winning team, in order to justify a claim to a share of the prize. However, in other cases these membership benefits strike us as objectionable and as grounding legitimate complaint from others—as in the case of providing tax breaks for legally married couples.[1] Following T. M. Scanlon, contractualists are interested in the personal claims of individuals. This focus means that contractualists must take seriously the questions of what counts as an individual's claim and whether individual claims are legitimate in cases in which they arise as a result of an individual's group-membership.

These questions, or so I argue, are especially difficult in cases where we must determine the permissibility of risk impositions. And taking these questions about individual claims seriously may lead us to rethink an approach to risk which is popular among contractualists, and the puzzle this approach gives rise to. So-called ex ante contractualism determines the claims of individuals by considering the probabilities of being harmed or benefitted which a principle of risk imposition imposes on individuals. But it also leads to a puzzle: ex ante contractualism delivers principles which license the imposition of large burdens on some, in order to secure smaller benefits for others, despite the fact that such principles stand in direct opposition to the individualist, anti-utilitarian spirit of contractualism.

---

[1] Once we exclude justifications based on perfectionist or traditionalist values, which attach value to marriage as an institution, or on instrumentalist concerns such as the incentivization for having children.

Jessica J. T. Fischer, *The Individualist Objection: or Why Ex Ante Probabilities Aren't Always Individualistic*
In: *Oxford Studies in Normative Ethics, Volume 12*. Edited by: Mark Timmons, Oxford University Press.
© Jessica J. T. Fischer 2022. DOI: 10.1093/oso/9780192868886.003.0005

This chapter provides a diagnosis of this puzzle: it argues that claims which cannot be justified on grounds other than by appeal to an individual's group-membership fail to constitute truly individual claims. Yet an individual's probability of being harmed or benefitted may be conditioned by an individual's group-membership—as when, for instance, an individual's probability of being selected for jury duty is conditioned by the size of the group from which jurors are chosen. This means that claims which appeal to these kinds of probabilities cannot be justified on grounds other than by appeal to an individual's group-membership and therefore fail to constitute truly individual claims. This, so the chapter concludes, is what gives rise to the puzzle of ex ante contractualism.

The chapter is structured as follows: the first section outlines the puzzle of ex ante contractualism, while the second section motivates the contractualist tenet of individualist justification. The third section argues that claims which are exclusively based on group-membership fail to adhere to individualist justification. Drawing on this analysis, the fourth section offers a diagnosis of the puzzle of ex ante contractualism.

1. The Puzzle of Ex Ante Contractualism

In what follows, I look at how contractualists deal with risk, and explain the puzzle of ex ante contractualism which at least some of them face.

Contractualism tells us that a moral principle is wrong if and only if it can be 'reasonably rejected' by individuals (Scanlon 1998: 85). A moral principle can be reasonably rejected by an individual if the complaint that the individual has against that principle is stronger than the complaint that any other individual has against any alternative principle. Contractualism also imposes a constraint on the types of complaints which individuals may advance by stating that individuals can only put forward complaints which are based on *personal grounds*, such as an individual's well-being, status, or rights. This concern for individualist justification bars individuals from advancing aggregative complaints which appeal to the summed-up complaints of several people. As such, the contractualist constraint of individualist justification provides a safeguard against maximizing principles which permit the imposition of large burdens on some in order to secure smaller benefits for others.[2] Individualist justification, then, lies at the core of contractualism's anti-utilitarian spirit.

But what can contractualism tell us about risk? Decision-making in medical, infrastructural, political, social, and various other spheres depends on having a framework for evaluating the permissibility of risk impositions. Yet in order to arrive at moral principles which regulate the permissibility of risk impositions, contractualists first need to settle on how to determine the complaints which individuals have in cases of risk. Here, contractualists are often urged to choose between assessing individual complaints from an *ex ante* or an *ex post* perspective. While ex ante contractualism evaluates the prospects which individuals have in light of the probabilities of harm which a principle imposes on an individual, ex post contractualism considers the complaints which individuals would be able to make against a principle, once an imposition of risk has played out.

Many find ex post contractualism to be attractive, since its concern for the most burdened individual aptly reflects a contractualist concern for the individual with the strongest claim. However, ex post contractualism also delivers intuitively troubling conclusions. Consider the following case:

Dangerous Treatment: 100 children are given an experimental treatment against a virus which will otherwise leave them paraplegic. Since the treatment imposes a 1/100 probability of death on each of them, one of them is likely to die as a result of the treatment, while the other 99 children will be cured.

Following ex post contractualism, *Dangerous Treatment* is impermissible, since the child who will die as a result of the treatment can reasonably reject it.[3] But this seems overly cautious. Plausibly, every child has reason to accept a very small, 1/100 probability of death in order to secure a very large, 99/100 probability of avoiding paraplegia.

A second, more damning objection to ex post contractualism points out that ex post contractualism prohibits many everyday social activities (Frick 2015).[4] Since activities such as driving or aviation render it highly likely that individuals will die as a result of them, ex post contractualism states that

---

[2] Individuals are equally excluded from appealing to impersonal values such as beauty.
[3] There is, of course, no absolute certainty that one of the children will die.
[4] As long as alternatives to it wouldn't result in an even greater number of deaths or harms.

these individuals can reasonably reject principles which license such activities.[5] Yet this seems clearly amiss.

Appealingly, ex ante contractualism can avoid both of these charges. First, ex ante contractualism rules *Dangerous Treatment* permissible because it is in each child's ex ante interest to secure a 99/100 probability of avoiding paraplegia at the cost of a 1/100 probability of death. Second, ex ante contractualism generally allows risky social activities such as driving, since each individual's probability of being harmed is minuscule, especially when compared to the benefits many people receive from these activities.

Intuitively attractive though ex ante contractualism may be, one forceful objection to it remains, which is the object of this chapter. Ex ante contractualism delivers implausible conclusions in cases such as the following:

Human Experiment: At t_1, a doctor organises a lottery among a group of 100 paraplegic children, which selects [one] of them by a random process. At t_2, he conducts medical experiments on [the selected child], foreseeing (but not intending) that [the child] will die in the course of the experiments. He knows for certain that the knowledge gained in this way will allow him to cure the remaining [99] children of their paraplegia. (Frick 2015: 204)[6]

Ex ante contractualism regards *Human Experiment* as structurally identical to *Dangerous Treatment*. It suggests that because each child's 1/100 probability of being experimented on is outweighed by their 99/100 probability of being cured, being part of the lottery is in the interest of each child. Thus, it rules *Human Experiment* permissible. Yet following our moral attitudes, *Human Experiment* seems wrong.

Two ways of defending ex ante contractualism have been offered. First, Johann Frick proposes that a principle must be justifiable not merely at the outset t_1 but at all other causally relevant ex ante stages of a procedure t_2, t_3, etc. (Frick 2015: 205). But since *Human Experiment* cannot be justified to the child who is chosen for experimentation at the causal stage t_2, *Human Experiment* is impermissible. Second, Rahul Kumar and Kerah Gordon-Solmon suggest that individuals have good reason to want to retain the 'decision-making authority over the use of [their] body' (Kumar 2015: 37).[7]

---

[5] For defences of ex post contractualism, see Reibetanz Moreau (1998) and Otsuka (2015).
[6] Number of children to be experimented on changed from 10 to 1.
[7] See also Gordon-Solmon (2019).

Since the medical experimentation taking place in *Human Experiment* violates this authority, *Human Experiment* can be reasonably rejected.

Both responses point to aspects of *Human Experiment* which are morally salient. But these aspects might well be contingent, and the impermissibility of *Human Experiment* might well be overdetermined. In other words, although both defences of ex ante contractualism are seemingly able to explain the impermissibility of *Human Experiment*, this doesn't entail that they have isolated the special wrong-making feature which unifies all cases in which ex ante contractualism produces problematic verdicts.[8] For instance, both defences cannot explain the following case:

Mountain Trip: High up in the mountains, Tilly will be crushed by a falling boulder, leaving her paraplegic. At t_1, you can throw a rock in the boulder's way which will redirect the boulder towards a far-away crevasse in which 1000 mountaineers are trapped. As the mountaineers are too tightly packed to move out of the boulder's pathway, the boulder will crush and kill one of them.

Tilly's complaint against certain paraplegia outweighs each mountaineer's complaint against a very small, 1/1000 probability of death. Thus, ex ante contractualism licenses *Mountain Trip* despite the fact that this delivers the implausible conclusion that the death of one mountaineer is preferable to Tilly's paraplegia. Yet since *Mountain Trip* neither contains separate causal stages nor the violation of an individual's decision-making authority over their body, both previous defences of ex ante contractualism fall short.

In addition, neither defence addresses the fact that *Human Experiment* and *Mountain Trip* are mere illustrations of a broader, structural feature of ex ante contractualism: because ex ante contractualism proceeds by comparing individual probabilities, it necessarily—mathematically—licenses the imposition of large burdens on some in order to secure smaller benefits for others in all cases in which both very low probabilities of being harmed and very high probabilities of being benefitted are imposed on individuals.[9] Yet this kind of excessive burdening of some, in order to secure smaller benefits for others, is reminiscent of the maximizing conclusions

---

[8] For another objection to Frick's response, see Gordon-Solmon (2019).
[9] This is also highlighted in Horton (2017).

utilitarianism delivers, and something a contractualist would reject (Scanlon 1998: 208–9).[10]

This leaves us with a puzzle. Given that contractualism deliberately sets out to exclude maximizing principles through individualist justification, how is it possible that a contractualist approach to risk winds up licensing the imposition of large burdens on some, in order to secure smaller benefits for others? Where has ex ante contractualism gone wrong? The rest of this chapter is dedicated to providing an answer to the diagnostic question of how the puzzle of ex ante contractualism arises in the first place.

2. Individualist Justification

The upcoming diagnosis of the puzzle of ex ante contractualism claims that ex ante contractualism mistakenly admits complaints which violate individualist justification. In order to be able to appreciate why this is deeply worrying for contractualists, it's important to expand on the motivation behind a contractualist concern for individualist justification.

Individualist justification stipulates that individuals can only advance complaints based on personal grounds and thus bars individuals from appealing to the aggregated claims of several individuals. The attraction of such a bar on aggregation is often illustrated through the following case:

Transmitter Room: Suppose that Jones has suffered an accident in the transmitter room of a television station…and we cannot rescue him without turning off the transmitter for fifteen minutes. A World Cup match is in progress, watched by many people, and it will not be over for an hour. Jones…is receiving extremely painful electrical shocks. (Scanlon 1998: 235)

If aggregation is permitted, each viewer can appeal to the displeasure which an interruption of the match will cause to millions of viewers, in order to outweigh Jones's complaint and to reasonably reject the interruption of the match. But this seems like the wrong conclusion. Barring aggregation delivers the more plausible verdict that because Jones has the strongest complaint, Jones can reasonably reject the continuation of the match.

---

[10] More recently, Scanlon (2013) has shown sympathy for ex ante contractualism.

Although many agree that aggregation is objectionable in cases like *Transmitter Room*, they also reject an absolute bar on aggregation as proscribed by individualist justification.[11] In what's referred to as *the numbers problem*, it's pointed out that in a rescue case in which one can save either one individual or two individuals from death, but cannot save all three, individualist justification permits us to save the single individual. But this conclusion, so the objection goes, is strongly implausible.[12]

With the numbers problem being beyond this chapter, the pertinent question is whether contractualists have independent reason for persevering with individualist justification, despite its potentially controversial upshots. Contractualists can respond by offering the following grounds for adopting individualist justification: moral principles must be justifiable to individuals based on personal grounds because we must be able to justify our actions *to them*. Thus, the process is one of not only individualist but mutual justification, in which principles are justified from one individual to another (Scanlon 2003: 434). This process is essential because it reflects both how individuals value each other as self-governing, reason-assessing creatures and how they value the mutual relationship in which they stand with one other. That is, for an individual to justify herself towards another person means to recognize the other as a rational, reason-assessing person with the same moral standing as herself, and to recognize the implications of this.[13]

Firstly, this involves recognizing that there are reasons to treat the other person in certain ways, such as reasons not to make them suffer, or reasons to protect them from destitution. Secondly, this requires recognizing that there are certain constraints on the grounds which an individual can permissibly appeal to when justifying themselves to another person. These constraints include an exclusion of reasons which cannot be justified on purely personal grounds, such as aggregative reasons. This is because, when appealing to aggregative reasons in order to justify themselves, the individual is failing to take seriously the personal claims of the opposing individual and the fact that the other individual has her own, separate reasons, and her

---

[11] Note that a requirement to save Jones in *Transmitter Room* can be accommodated by affirming an account of limited aggregation instead of an absolute bar on aggregation. For such an approach, see Voorhoeve (2014).
[12] For the first articulation of this objection, see Parfit (2003). For Scanlon's response, see Scanlon (1998: 232–4). For further discussion of Scanlon's response, see Otsuka (2006) and Munoz-Dardé (2005).
[13] For further discussion of mutual justification, see Kumar (2012).

own life to live. As such, the individual fails to value the other as a reason-assessing individual who is owed justification.

3. Collective Complaints

Having seen why individualist justification bars aggregative complaints, this section proposes that there is a second type of complaint which equally violates individualist justification. Although more elusive than aggregative complaints, complaints which are exclusively justified by appeal to an individual's group-membership also cannot be justified on personal grounds.

How can complaints fail to be justifiable on personal grounds without being aggregative? To give but a rough outline, consider the following: when advancing an aggregative complaint, an individual is adding the claims of other people to their own claim, and then appealing to the resulting sum. Thus, aggregative complaints *explicitly* rely on the claims of other people. In comparison, when presenting what we may call a *collective complaint*, an individual is advancing a complaint which cannot be justified on grounds other than by appeal to an individual's membership in a group with others. Collective complaints therefore *implicitly* rely on the claims of other people.

A deeper understanding of collective complaints can be gained by outlining their two key features: first, collective complaints must include an appeal to a *collective property*. Collective properties are properties which can only be held collectively, and which cannot be held by individuals themselves. Examples of collective properties include being married, being a team member, being a co-habitant, being a co-plaintiff, or being a fellow countryman. Each of these is a collective property since, in the absence of others and in the absence of the respective collective, an individual cannot possibly be married, be a team member, co-habitant, co-plaintiff, or fellow countryman. Second, collective complaints must have no independent personal justification: if a complaint can be justified by an appeal to personal grounds alongside its appeal to a collective property, it doesn't constitute a collective complaint.

What renders collective complaints truly collective is found in the second feature: collective complaints cannot be justified on grounds other than by appeal to a collective property. But collective properties are constituted by the presence of other people, and therefore cannot be justified based on personal grounds. Accordingly, collective complaints cannot be justified on personal grounds and violate individualist justification. Attention to individualist justification, then, should lead contractualists to bar individuals from advancing collective complaints.

Let's consider some cases. First, it's helpful to look at a case in which a complaint fails to constitute a collective complaint, because it only fulfils the first but not the second feature.

Relay Team: A community decides whether to discontinue the small financial benefit awarded to each member of the local relay team. Together with Mac and Milly, Mia has been running for the local relay team. Mia is unhappy about the proposed change.

Mia's complaint against the repeal of team benefits fulfils the first feature of collective complaints since it appeals to the collective property of Mia's membership in the relay team. Yet Mia's complaint fails to fulfil the second feature of collective complaints because it can still be justified on personal grounds. In particular, Mia's complaint can be justified by appeal to Mia's personal contributions to the relay team: Mia has attended practice, run races, visited events, etc., and thus earned her benefits. Because Mia's complaint in *Relay Team* can be justified on such personal grounds, it doesn't constitute a collective complaint.

Keeping this in mind, we can move on to a case which contains a collective complaint:

Council Tax Breaks: A government decides whether to repeal council tax breaks for households with multiple occupants. Mia, who lives in a two-person household, is unhappy.

We need to be careful here. There is a question about whether Mia has a complaint based on the fact that she will lose out financially as a result of *Council Tax Breaks*. But what I'm interested in is the question of whether Mia may appeal to the fact that she is part of a two-person household in order to advance a complaint against *Council Tax Breaks*, which states that she is deprived of something she has a claim to.

Note that a complaint of this kind appeals to the fact that Mia is a cohabitant in a two-person household, and thus appeals a collective property: just by herself, Mia cannot be a co-habitant, just like Mia, just by herself, cannot be the sole member of the relay team. Yet unlike in *Relay Team*, Mia's complaint fulfils the second feature of collective complaints. Mia's complaint in *Council Tax Breaks* cannot be justified on grounds other than by appeal to Mia's collective property of being a co-habitant. Thus, Mia's complaint is collective. As a result, a concern for individualist justification should lead contractualists to bar Mia from advancing her complaint, just as contractualists would bar her from advancing an aggregative complaint.

There are ways of understanding this analysis of collective complaints which might seem attractive, but which are based on various misinterpretations. Spelling out how three of these ways go awry allows us to gain a richer understanding of collective complaints.

A first misreading might invoke a popular and widely accepted definition of individualist justification in order to reject the idea that collective complaints violate individualist justification:

Individualist Restriction: In rejecting some moral principle, we must appeal to this principle's implications only for ourselves and for other *single* people. (Parfit 2011: 193)

Introduced by Derek Parfit, this definition of individualist justification bars individuals from advancing aggregative complaints because aggregative complaints appeal to the implications which a principle has for other people. Collective complaints, however, only appeal to the implications which a principle has for the individual herself and, therefore, fail to violate the individualist restriction.

But this proposal overlooks that we have at least three reasons for rejecting this definition of individualist justification. First of all, contractualists should be wary of cashing out individualist justification in terms of a principle's implications or outcomes, given that they ultimately care about recognizing individuals, and about the relationships that hold between them. Such an interpretation of individualist justification not only seems potentially coloured by Parfit's consequentialist sympathies. It also seems inappropriate in light of the task at hand, which consists of identifying the core tenet of a moral theory constructed with the purpose of eschewing a utilitarian, outcome-centric approach.

Second, contrary to what the individualist restriction as offered by Parfit may indicate, a bar on collective complaints aligns with Scanlon's own discussion of individualist justification. Scanlon distinguishes between two types of complaint: 'reasons arising from the points of view of individuals' and 'reasons corresponding to the claims of groups of individuals' (Scanlon 1998: 231). And while Scanlon discusses aggregative reasons as belonging to the latter group, we have no grounds for presuming that complaints 'corresponding to the claims of groups of individuals' are exhausted by aggregative reasons. In fact, given that collective complaints are exclusively justified by appeal to a collective property such as group-membership, collective complaints straightforwardly 'correspond to the claims of groups of individuals'.

Third, collective complaints fall within the concerns which originally motivate the constraint of individualist justification, as they were outlined in the previous section. This is because, like aggregative complaints, collective complaints exclusively arise from the viewpoint of a collective and thus fail to arise 'from the points of view of individuals' (Scanlon 1998: 231). But since principles are justified to individuals rather than collectives (or individuals qua members of collectives), advancing a complaint which exclusively arises from the viewpoint of a collective fails to recognize and respect the person to whom one is offering justification. In other words, just like aggregative complaints, collective complaints fail to take seriously the personal claim of the other individual. Therefore, collective complaints impede the mutual justification of principles between individuals and violate individualist justification.

The proposal at hand might be misinterpreted in a second way. One might claim that a bar on collective complaints brings along the exclusion of certain other individual complaints which contractualists would be adamant to keep, such as, for instance, complaints against being arbitrarily favoured or disfavoured. Consider the following case:

Unequal Pay: Milly is arbitrarily paid less than her colleagues.

One might suggest that Milly's complaint against being paid less than her colleagues cannot be justified on grounds other than by appeal to the collective property of Milly's membership in her group of colleagues.[14] Thus, Milly's complaint is collective and violates individualist justification.

Let me show why Milly's complaint isn't collective. While Milly's complaint depends on the collective property of her membership in her group of colleagues and thus fulfils the first feature of collective complaints, it doesn't fulfil the second one: although downstream from Milly's membership in a group of colleagues, the relevant property in *Unequal Pay* refers to the fact

---

[14] I am indebted to Jay Wallace for this point.

that Milly is paid less than her colleagues. But 'being paid less than one's colleagues' isn't a collective property, since it isn't held by Milly's colleagues. Thus, Milly's complaint in *Unequal Pay* can be justified on personal grounds, namely by appeal to Milly's personal interest in not being disfavoured arbitrarily.

This brings out a general point: many individual complaints rely on collective practices or collective endeavours and thus rely on collective properties. We only need to think of the complaints individuals have in contexts such as promise-breaking or the provision of social goods. But while such individual complaints clearly supervene on collective properties, they can generally be justified on personal grounds downstream from these properties. By contrast, what I'm worried about is cases in which complaints don't only supervene on a collective property but are themselves collective in so far as they can be reduced to an appeal to the relevant collective property. Such complaints fail to be justifiable on personal grounds and therefore must be excluded by thoroughgoing contractualists who take the constraint of individualist justification seriously.

In a third misunderstanding, someone might dismiss the proposal on collective complaints as a moot point. They might draw attention to the fact that in cases like *Council Tax Breaks*, it is irrelevant whether or not Mia has a complaint based on the fact that she lives in a two-person household. This is because even if Mia should have such a complaint, it will be outweighed by the complaints based on arbitrary disfavouring which all individuals in single-person households have against such unfair tax breaks. In effect, collective complaints will standardly be outweighed by conflicting complaints, as individuals with the latter can reasonably reject principles which arbitrarily disfavour them based on collective properties. Therefore, the category of collective complaints is superfluous.

While it might be true that collective complaints are oftentimes outweighed, the category of collective complaints remains normatively significant, for three reasons. First, collective complaints advantage individuals who are part of collectives—such as those living in multiple-person households—over individuals who aren't—such as those living in single-person households—and thus technically allow individuals to reap benefits purely based on their group-membership. Consequently, allowing individuals to advance collective complaints assigns normative relevance to brute group-membership, and technically promotes principles which permit the burdening of some in order to provide smaller benefits to others. For contractualists who set out to controvert maximizing principles of such ilk by adopting individualist justification, these seem like egregious affirmations to make.

Second, the concept of collective complaints finds applicability in unusual cases. Specifically, it's helpful in cashing out complaints which fail to be aggregative, but which aren't fully personal either. Consider the following example: responding to the numbers problem, Scanlon proposes that contractualists can arrive at a duty to save the greater number without aggregating claims. Scanlon claims that when choosing between saving one individual or two individuals, the presence of the second individual in the larger group breaks the tie between both groups, since the individual otherwise has a complaint against not being recognized (Scanlon 1998: 232–4). Michael Otsuka rejects Scanlon's argument by highlighting that the second individual can only break the tie between both groups if their claim is considered *together* with the first individual's claim. This, so Otsuka points out, seems to violate individualist justification (Otsuka 2000: 291). Yet, evidently, Scanlon's argument doesn't employ aggregation. So, what's going on here? Equipped with the concept of collective complaints, we can explain that the second person's complaint against not being recognized violates individualist justification because it advances a collective complaint: it cannot be justified on grounds other than by appeal to the collective property of being a member in the group of two individuals. In brief, the concept of collective complaints allows contractualists to make sense of complaints which non-aggregately violate individualist justification, and to reflect them from within their own framework.

Third, the category of collective complaints turns out to be paramount for the puzzle of ex ante contractualism. This is because many complaints which appeal to an individual's ex ante probabilities are in fact collective complaints, and thus barred by individualist justification. The remainder of this chapter focuses on establishing this point.

4. Diagnosing the Puzzle of Ex Ante Contractualism

Let's return to our initial puzzle: we've been wondering why ex ante contractualism winds up delivering principles which license the burdening of some in order to secure smaller benefits for others, despite the fact that contractualism specifically sets out to avoid such maximizing principles. This puzzle arises in cases in which both very low probabilities of being harmed and very high probabilities of being benefitted are imposed on individuals—as in *Human Experiment*.

In this section, a diagnosis of the puzzle of ex ante contractualism is outlined: I explain that, oftentimes, complaints which appeal to an individual's ex ante probabilities of being harmed or benefitted are in fact collective complaints and therefore violate individualist justification. This allows us to see that it is the admission of collective complaints which are in violation of individualist justification that causes ex ante contractualism to counter-intuitively license principles which impose large burdens on some in order to secure smaller benefits for others.

To begin this outline, it's helpful to reflect upon a feature of ex ante contractualism which is often taken as a strength of this view. Following ex ante contractualism, an individual's ex ante probabilities may rise or fall according to the number of people affected by a principle. For instance, an individual's ex ante probability of being called for jury duty is higher if the population pool from which jurors are selected is smaller and lower if the population pool from which jurors are selected is larger. In other words, there is a sense in which ex ante contractualism is indeed sensitive to the number of people. All the while, we might have thought it impossible to count the numbers without aggregating, and without violating individualist justification. Yet it seems as if ex ante contractualism succeeds at this feat: because an individual's probability of being harmed or benefitted by a principle only appeals to the individual's well-being, a complaint which appeals to such probabilities remains individualist in its justification, while taking the number of individuals into consideration (Frick 2015: 201). Striking this kind of balance might be thought a rather fine achievement for contractualists.

Still, one might also take a different perspective. One might wonder whether the fact that ex ante contractualism is sensitive to the number of people signals that there is something amiss with ex ante contractualism which has so far escaped our notice. Note that ex ante contractualism can count the numbers due to the fact that it allows for an individual's ex ante probabilities to correspond to the size of a population. Yet such correspondence seems highly pertinent to the puzzle of ex ante contractualism. After all, each child's probability of death in *Human Experiment* is only 1/100 because they are part of a population of precisely 100 participating children. With that said, the way in which ex ante probabilities may correspond to population size requires closer scrutiny.

Let's tease out in more general terms how an individual's probabilities sometimes correspond, and sometimes fail to correspond, to population size. I suggest that individual probabilities are either *number-dependent* or *number-independent*. Individual probabilities are number-independent if they in no way correspond to the number of individuals in a population. Consider the following example:

Lightning Strike: On average, an individual has a 1/1000 probability of being hit by lightning in any given year.

In *Lightning Strike*, an individual's 1/1000 probability of being hit by lightning is entirely independent of the number of other individuals who are exposed to the same risk. Whether the individual is the only person on earth, or one of billions of other people, their probability of being hit by lightning remains the same. Thus, in *Lightning Strike*, an individual's 1/1000 probability of being hit by lightning is number-independent. Instead, the individual's probability depends on alternative factors, such as the physical prevalence of lightning.

By contrast, individual probabilities are number-dependent if they necessarily correspond to the number of individuals in a population in one out of two ways. First, individual probabilities are number-dependent if they refer to a fixed number out of a population. Consider the following example:

Councillor Duty: In a village, 10 councillors are chosen by lot every year, and the village has an eligible population of 1000. Thus, each eligible individual has a 10/1000 probability of being selected as councillor every year.

In *Councillor Duty*, each individual's probability of being selected as councillor changes if either the number of required councillors or the population size changes. If, for instance, the town's population doubles, each individual's probability of being selected as councillor decreases from 10/1000 to 10/2000. This is because the numerator of each individual's 10/1000 probability is the number of individuals on whom the burden of being councillor will be imposed, and the denominator is the population size out of which those individuals are chosen. Thus, each individual's probability necessarily corresponds to the number of individuals on whom a burden will be imposed and to the number of individuals in the population, rendering it number-dependent.

Second, individual probabilities are also number-dependent if they refer to a fixed fraction out of a population:

Jury Duty: In a large city, 10 jurors are required for every 1000 inhabitants in order to fill judiciary needs. Thus, each eligible individual has a 10/1000 probability of being selected for jury duty in a year.

In *Jury Duty*, each individual's 10/1000 probability of being selected for jury duty consists of a fixed fraction out of a population. This fraction remains the same, seemingly independent of the number of individuals in the city's population: whether the city has thousands or millions of inhabitants, each individual's probability to be selected for jury duty remains 10/1000.

But this merely obscures a more subtle way in which individual probabilities are number-dependent in *Jury Duty*. Note that each individual's 10/1000 probability of being selected depends on there being a need for jurors to regulate civil life in the first place, and on every individual's membership in the city's population of at least 1000 individuals. This is clear from the fact that once the city's population shrinks below 1000 individuals, each individual's probability becomes a fixed number probability which has the number of individuals required as jurors as numerator and the population size as denominator. For instance, should the city's population shrink to 900, each individual's probability of being selected for jury duty rises from 10/1000 to 10/900, at least if we assume that the number of required jurors remains the same. Thus, each individual's 10/1000 probability of being selected for jury duty is number-dependent, since it corresponds to the number of individuals in the population being 1000 or higher.

This breakdown provides us with a better understanding of how probabilities may correspond to population size, and, by extension, with insight into how ex ante contractualism succeeds at counting the numbers without aggregating: simply put, ex ante contractualism works with complaints which, in virtue of appealing to number-dependent probabilities, directly reflect the number of affected individuals. Crucially, the distinction between number-independent and number-dependent probabilities—with the latter corresponding to population size—presents us with a new angle from which to analyse the puzzle of ex ante contractualism.

Let's start by looking at *Dangerous Treatment*, the case in which ex ante contractualism provides plausible conclusions. In *Dangerous Treatment*, each child has a 1/100 probability of death from receiving an experimental treatment against a virus which will otherwise leave them paraplegic. But just as in *Lightning Strike*, this probability is entirely independent of the number of people affected. Whether the experimental treatment is given to 1, 5, 100, or 1,000,000 children, each child's 1/100 probability of death remains the same. This is because each child's 1/100 probability of death depends on medical factors regarding the nature of the treatment and is unaffected by population size. Thus, probabilities are number-independent in this case.

By contrast, ex ante contractualism delivers implausible conclusions in *Human Experiment*. On closer view, we can see that probabilities are number-dependent in this case, given that each child's 1/100 probability of death necessarily corresponds to the number of other children affected alongside them.[15] Given that the current description of *Human Experiment* underspecifies the case, it remains open in which of the two ways outlined probabilities are number-dependent.

On the one hand, individual probabilities in *Human Experiment* are fixed number probabilities analogous to *Councillor Duty* if it's stipulated that the doctor's experimentation on one child will lead him to discover the formula for a curative treatment against paraplegia, which allows him to cure any number of children, be it 99, 200, or 1000 children. In this scenario, should 1000 rather than 100 paraplegic children be affected, each child's probability of being experimented on falls to 1/1000. Either way, we can see that each child's probability has the number of experimented children as its numerator and, crucially, the population size as its denominator.

On the other hand, individual probabilities in *Human Experiment* are fixed fraction probabilities analogous to *Jury Duty* if we assume that experimentation on one child allows the doctor to produce a treatment which allows him to cure exactly 99 children from their paraplegia. Should 1000 rather than 100 paraplegic children be affected, the doctor would have to experiment on 10 children rather than one child in order to cure everyone. Thus, each child's 1/100 probability would remain constant even if the number of paraplegic children increases. Nevertheless, if the number of paraplegic children falls below 100, each child's probability becomes a fixed number probability, with the number of experimented children as its numerator and the population size as its denominator. Should, for instance,

---

[15] This also applies to *Mountain Trip*: each mountaineer's 1/1000 probability of death is number-dependent because it corresponds to the 1000 trapped mountaineers. It is only together that the 1000 mountaineers take up all the possible spaces which the boulder could land in, thus ensuring that one mountaineer will be killed.

only 90 paraplegic children be affected, each child's probability would rise from 1/100 to 1/90. Thus, each child's 1/100 probability in *Human Experiment* depends on their membership in a group of at least 100 paraplegic children.

This allows us to get a more detailed picture of the technical considerations which lead ex ante contractualism to deliver maximizing principles in cases in which both very low probabilities of being harmed and very high probabilities of being benefitted are imposed on individuals. Number-dependent probabilities decrease an individual's ex ante probabilities of being severely burdened by increasing the number of people across whom such a risk is spread. This occurs until an individual's probabilities against being severely burdened are small enough to be outweighed by competing complaints against lesser burdens. This is evident in *Human Experiment*, where every child's ex ante complaint against death is decreased to a complaint against a 1/100 probability of death due to the fact that the risk is spread across 100 children. As a result, consideration of number-dependent probabilities causes ex ante contractualism to license the imposition of large burdens on some in order to secure smaller benefits for others in all cases in which both very low probabilities of being harmed and very high probabilities of being benefitted are imposed on individuals.

Let's put all these strands together. Contractualism's tool for excluding maximizing principles is found in the constraint of individualist justification, which states that individuals may only advance complaints which are based on personal grounds. This anti-utilitarian constraint is violated by both aggregative and collective complaints. Now, it was already said that complaints which correspond to population size through relying on number-dependent probabilities can avoid the charge of aggregation. But they cannot avoid the charge of constituting collective complaints. As seen before, complaints are collective if, first, they appeal to a collective property and, second, they cannot be justified on grounds other than by appeal to a collective property. In cases in which complaints appeal to number-dependent probabilities, both of these features are generally fulfilled: first, number-dependent probabilities are a collective property, since, just like being a team member, number-dependent probabilities only arise collectively and are only held by individuals in virtue of their membership in each respective population. Second, individual complaints which appeal to number-dependent probabilities in cases like *Human Experiment* cannot be justified on personal grounds beyond an appeal to such probabilities.[16] We can now provide a diagnosis of the puzzle of ex ante contractualism. Why does ex ante contractualism wind up licensing the imposition of large burdens on some in order to secure smaller benefits for others, despite the fact that contractualism specifically sets out to avoid such principles? Answer: because by allowing individuals to advance collective complaints, it impermissibly proceeds via an avenue which is closed off by individualist justification.[17] In effect, this is analogous to how contractualism winds up licensing the imposition of a large burden on Jones in order to secure smaller benefits for the scores of TV viewers if it violates individualist justification and admits aggregative complaints in *Transmitter Room*.

This diagnosis has substantial implications for ex ante contractualism.[18] It entails that it is impermissible for contractualists to determine individual complaints by reference to probabilities if the latter are number-dependent, just as it's impermissible for contractualists to determine individual complaints in *Transmitter Room* by employing aggregation. Thus, ex ante contractualism, in its standard form, cannot be used to determine individual complaints in risk cases with number-dependent probabilities.[19]

---

[16] Complaints which appeal to number-dependent probabilities may sometimes fail to fulfil the second feature of collective complaints because they can be justified on personal grounds. This could be, for instance, a personal reason against being disadvantaged arbitrarily.

[17] Rawls and Scanlon both suggest that when individuals are choosing moral principles from behind a veil of ignorance, an appeal to ex ante probabilities is impermissible. The current diagnosis may well be used to provide support for this Rawlsian and Scanlonian exclusion of probabilities. See Rawls (1971) and Scanlon (1982). For a recent challenge to Rawls's claim, see Buchak (2017).

[18] This account also has implications for the numbers problem: some have proposed that one ought to run a weighted lottery in rescue cases, which assigns proportional chances of being saved to both groups (see, for instance, Kamm (1998), Saunders (2009), and Timmerman (2004)). In a rescue case in which one could either save one individual or two individuals, but cannot save all three, a weighted lottery assigns a 1/3 chance of being saved to the single individual and a 2/3 chance of being saved to the two individuals. Yet while one might think that the fact that a weighted lottery straightforwardly reflects the numbers of individuals is an advantage of this approach, the previous discussion reveals that it is not: because a weighted lottery only achieves this feat by relying on number-dependent probabilities, it fails to be justifiable on personal grounds and violates individualist justification. Therefore, contractualists should reject the running of a weighted lottery in rescue cases.

[19] There is a worry that suggests that ex ante contractualism delivers problematic conclusions not because it relies on number-dependent probabilities, but because it relies on probabilities at all. I am disinclined to grant that worry, but even if it were granted the following remains true: because ex ante contractualism admits collective complaints, it violates individualist justification and thus fails to present a thoroughly contractualist view. I am grateful to Alex Voorhoeve and Johann Frick for pointing me to this worry.

Let's quickly sketch and reject three ways in which this diagnosis of the puzzle of ex ante contractualism might be viewed as delivering implausible conclusions. First, the proposed diagnosis might be regarded as delivering the wrong conclusion in the following case:

Spoilt Treatment: 100 children are given an experimental treatment against a virus which will otherwise leave them paraplegic. Due to a storage accident, one dose has been rendered toxic. There is no way of telling which dose has been spoilt and no way to secure further doses. Thus, one of the children will die from receiving the toxic dose, while the other 99 children will be cured.

Following the previous discussion, each child's 99/100 probability of being cured appears to be number-dependent in *Spoilt Treatment*. This is because each child's probability of death is dependent on the presence of the 99 other children: the higher the number of children who have already received non-toxic doses, the higher each remaining individual's probability of receiving the toxic dose. If probabilities in *Spoilt Treatment* are number-dependent, this renders complaints in *Spoilt Treatment* inadmissible. Yet this verdict seems problematic: just as in *Dangerous Treatment*, it seems permissible to impose a very small probability of death on each child in order to secure a very large probability of avoiding paraplegia.[20]

This worry can quickly be dispelled. Appearances notwithstanding, probabilities in *Spoilt Treatment* aren't number-dependent. Each child's 1/100 probability of death isn't a collective property because it doesn't depend on each child's membership in the group of 100 children. This is evident from the fact that even if only 1, 10, or 50 children receive doses of the experimental treatment, each child's probability of death remains at 1/100. In similar vein as *Dangerous Treatment*, each individual's 1/100 probability of death arises as a result of alternative factors.

Second, one may suggest that the probabilities of harm which are imposed on individuals by risky social activities such as driving are number-dependent, since they roughly correspond to population size. This is evident from the fact that one stands a greater chance of being harmed when driving in a densely populated area than when driving in a sparsely populated area. If correct, this means that the present proposal bars individuals from appealing to the very low probabilities imposed by social

---

[20] For this case and for discussion of it, I'm grateful to Alec Walen.

activities such as driving in order to agree to principles which permit such activities.

This problem, too, is unsubstantiated. On closer view, probabilities of harm imposed by social activities such as driving fail to be number-dependent. This is because population size is just one among several causal factors which determine individual probabilities. Other, more significant causal factors include the technical safety of cars, local traffic laws, speed travelled at, urban planning, etc. Simply put, in cases such as *Human Experiment*, each child's probability of death is only low enough to make participation in the lottery plausible for each child because it is decreased by the fact that a sufficiently large number of other children are participating alongside them. But the low probability of harm which is imposed on individuals through social activities such as driving is not a product of risk dispersal but a product of technological and infrastructural factors, such as road safety. After all, this is why even the only inhabitant of an abandoned town exposes themselves to a risk of death while driving.

Third, following a bar on complaints which appeal to number-dependent probabilities, it may seem as if *Human Experiment* cannot be reasonably rejected by individuals. Yet if *Human Experiment* cannot be reasonably rejected and is therefore permissible, this delivers precisely the verdict which we set out to avoid.

This point is based on a misunderstanding. The present diagnosis of the puzzle of ex ante contractualism suggests that because complaints which appeal to number-dependent probabilities violate individualist justification, they are inadmissible according to contractualist moral theory. This doesn't convey a negative moral verdict but conveys no moral verdict at all. As such, the present diagnosis doesn't rule *Human Experiment* permissible. It simply states that if all we have to go on in a case are collective complaints, then contractualism cannot admit these complaints and, therefore, is unable to reach any moral verdict.

Still, one question remains: in light of these implications, how should contractualists assess cases of risk? Should they opt for a hybrid approach which applies ex ante contractualism in some cases and ex post contractualism in others?[21] Or adopt ex post contractualism after all? While answering the question of how contractualists should respond to risk is a project for another time, one might consider this discussion as impetus for challenging

---

[21] For a hybrid account, see Walen (2020).

the sharp boundary between ex ante and ex post approaches. Such a challenge might be pursued by re-evaluating Scanlon's own proposal that risk impositions are permissible once 'reasonable precautions' have been taken (Scanlon 1998: 209). Scanlon suggests that an individual's probability of being harmed might inform our reasoning, not by constituting and thus diminishing an individual's complaint, but by indicating what level of care agents must apply when proceeding with a risky action. Such precautions, we might think, simply aren't taken in cases like *Human Experiment*.

To summarize, this section has argued that ex ante contractualism mistakenly admits collective complaints into its considerations, despite the fact that such complaints are excluded by individualist justification. This mistake has had two major upshots. First, an apparent virtue: the admission of complaints which rely on number-dependent probabilities allows ex ante contractualism to count the numbers without aggregating. On the flipside of this virtue, admitting complaints which rely on number-dependent probabilities licenses the imposition of large burdens on some in order to secure smaller benefits for others, and thus spawns the puzzle of ex ante contractualism.

Conclusion

The main aim of this chapter was to provide an analysis of the puzzle of ex ante contractualism and to explain why ex ante contractualism may possibly deliver maximizing principles despite the fact that such principles stand in direct opposition to the anti-utilitarian spirit of contractualism.

The kind of reasons moral theorists may appeal to depend on their larger theoretical commitments. Thoroughgoing contractualists, or so I have argued, must only be concerned with complaints which can be justified to individuals on personal grounds, and therefore exclude group-based, collective complaints from their reasonings. It is because ex ante contractualism—in its standard form—fails to exclude collective complaints, and instead relies on them to arrive at moral principles, that it falls short of providing a purely contractualist approach to risk. According to the diagnosis of this chapter, then, misunderstandings of the nature of individualist justification and of probabilities have obscured that there is something amiss with ex ante contractualism: by allowing individuals to advance collective complaints, ex ante contractualism violates the core contractualist tenet of individualist justification.[22]

References

Buchak, Lara. 2017. "Taking Risks behind the Veil of Ignorance." *Ethics* 127: 610–44.

Frick, Johann. 2015. "Contractualism and Social Risk." *Philosophy & Public Affairs* 43: 175–223.

Gordon-Solmon, Kerah. 2019. "Should Contractualists Decompose?" *Philosophy & Public Affairs* 47: 259–87.

Horton, Joe. 2017. "Aggregation, Complaints, and Risk." *Philosophy & Public Affairs* 45: 54–81.

Kamm, Frances. 1998. *Morality, Mortality: Death and Whom to Save from It.* Oxford: Oxford University Press.

Kumar, Rahul. 2012. "Contractualism on the Shoal of Aggregation." In R. Jay Wallace, Rahul Kumar, and Samuel Freeman, eds., *Reasons and Recognition: Essays on the Philosophy of T. M. Scanlon*, 129–55. Oxford: Oxford University Press.

Kumar, Rahul. 2015. "Risking and Wronging." *Philosophy & Public Affairs* 43: 27–51.

Munoz-Dardé, Véronique. 2005. "The distribution of numbers and the comprehensiveness of reasons." *Proceedings of the Aristotelian Society* 105: 191–217.

Otsuka, Michael. 2000. "Scanlon and the Claims of the Many versus the one." *Analysis* 60: 288–93.

Otsuka, Michael. 2006. "Saving Lives, Moral Theory, and the Claims of Individuals." *Philosophy & Public Affairs* 34: 109–35.

Otsuka, Michael. 2015. "Risking Life and Limb." In I. Glenn Cohen, Norman Daniels, and Nir Eyal, eds., *Identified versus Statistical Lives*, 77–93. Oxford: Oxford University Press.

Parfit, Derek. 2003. "Justifiability to Each Person." *Ratio* 16: 368–90.

[22] For helpful comments and conversations, I'm grateful to Sebastien Bishop, Luke Fenton-Glynn, Johann Frick, Joe Horton, Kacper Kowalczyk, Véronique Munoz-Dardé, Bastian Steuwer, Alice van't Hoff, Nikhil Venkatesh, Alex Voorhoeve, Alec Walen, Jay Wallace, and Han van Wietmarschen. I'm also grateful to audiences at the 2021 Arizona Workshop in Normative Ethics, the UCL moral and political philosophy group, and the Princeton Ethics group.

Parfit, Derek. 2011. *On What Matters*. Oxford: Oxford University Press.

Rawls, John. 1971. *A Theory of Justice*. Cambridge: Harvard University Press.

Reibetanz Moreau, Sophia. 1998. "Contractualism and Aggregation." *Ethics* 108: 296–311.

Saunders, Ben. 2009. "A Defence of Weighted Lotteries in Life Saving Cases." *Ethical Theory and Moral Practice* 12: 279–90.

Scanlon, T. M. 1982. "Contractualism and Utilitarianism." In Amartya Sen and Bernard Williams, eds., *Utilitarianism and Beyond*, 103–28. Cambridge: Cambridge University Press.

Scanlon, T. M. 1998. *What We Owe to Each Other*. Cambridge: Harvard University Press.

Scanlon, T. M. 2003. "Replies." *Ratio* 16: 424–39.

Scanlon, T. M. 2013. "Reply to Zofia Stemplowska." *Journal of Moral Philosophy* 10: 508–14.

Timmerman, Jens. 2004. "The Individualist Lottery: How People Count but not Their Numbers." *Analysis* 64: 106–112.

Voorhoeve, Alex. 2014. "How Should We Aggregate Competing Claims?" *Ethics* 125: 64–87.

Walen, Alec. 2020. "Risks and Weak Aggregation: Why Different Models of Risk Suit Different Types of Cases." *Ethics* 131: 62–86.

5

Blessed Lives, Bright Prospects, Incomplete Orderings

Jamie Dreier

1. Preliminaries

Better

Some ways the world could be are better than others. It's better if the two candidates address the issues and don't interrupt each other; it's better for you if you don't get infected with a lethal virus. Some people are skeptical about the idea of an outcome being just Better[1]; but in this chapter I'll assume it makes sense. Which outcomes are Better is, typically, a function of how good they are for the persons affected. Maybe not when there are morally significant non-persons involved, or when some options are fairer than others. But no such complications are involved in the central cases I'm going to be considering.[2]

Here is a very plausible principle, the "Principle of Personal Good" (PPG), due to Broome (1995):

(PPG)   If one option is better for someone and at least as good for everyone, then it is Better.

The PPG is a principle of *separability* in the dimension of persons. That is, it is a kind of dominance principle, and it can also be thought of as insisting that all of the good in a distribution is located in separate persons. I'll say more about separability in Section 2.

---

[1] I'll capitalize 'Better' for the general comparative, leaving it lower-case for the comparative for particular persons. Examples of skeptics about general Better include Thomson (1997) and Geach (1956).
[2] I'll have an example involving fairness below, but only to illustrate the significance of separability, not to make any polemical point.

Jamie Dreier, *Blessed Lives, Bright Prospects, Incomplete Orderings* In: *Oxford Studies in Normative Ethics, Volume 12.* Edited by: Mark Timmons, Oxford University Press. © Jamie Dreier 2022.
DOI: 10.1093/oso/9780192868886.003.0006

Incomparability and Parity

Sometimes it strikes us that neither of a pair of alternatives is better than the other, but not because they are equally good. These are cases of apparent incomparability.[3]

The hallmark of this phenomenon is resistance to small improvements. Suppose Mark could win as a prize a romantic Alaska vacation, or a trip to Kaliningrad to visit Kant's home. We might decide that neither is better for him. Maybe that's because we think the two trips are equally good for him, exactly balanced. But, according to many people, that is not very plausible. For if they were exactly balanced, then if it turned out that the Kaliningrad trip came with €20 in pocket money, so it's just a *little* bit better than we had thought, that wouldn't make it better than the Alaska vacation. We're still inclined to regard each as being no better or worse than the other.

Some people think that, at least sometimes, what is happening in cases of apparent incomparability is that one outcome is no Better (or no better for you) than a second, and no Worse; nor are they equally Good. They are, as Chang (2002) says, *on a par*. Parity is not vagueness, and it is not indeterminacy. When two consequences are on a par, it is fully determinate that neither is at least as good as the other: parity is just another relation that can obtain between them, exclusive of the ordinary ordering relations. Indeed, taking 'better' as primitive, we could define other value comparatives like this:

x is *worse* than y iff y is better than x

x is *exactly as good* as y iff (∀z) (x is better/worse than z iff y is better/worse than z)

x is *on a par* with y iff neither is better than nor exactly as good as the other.[4]

---

[3] Thanks to Leo Yan for the terminology. Incomparability is one diagnosis of the phenomenon, but it's better not to assume from the outset that the apparently incomparable items really are incomparable.

[4] This is not Chang's conception of *on a par*. For Chang, options have to be both comparable to some one thing, or they aren't on a par at all. For instance, Santana's album *Supernatural* isn't on a par with home-grown tomato salad, because even though neither is aesthetically better than the other, there isn't (at least, not obviously!) anything they're both better than, or that one of them is better than and the other worse than. There does seem to be an important difference between the kind of 'rough comparables' that fuel Chang's theory and pairs that are completely incomparable, but I don't think the difference will be confusing in our context, nor will it exert any rational pressure on verdicts or arguments here.

For most of the chapter, the interpretation of apparent incomparability that I will be working with is parity. In Section 5 I'll return to the other most likely interpretations: indeterminacy and ignorance.

The Goal

Separability principles, like the PPG, allow us to explain why one distribution is better than another on the basis of how good each is for the persons to whom the goods are distributed. This is a general feature of separability, as I'll spell out in Section 2. I want to know how the parts and elements of options contribute to the explanation of which options are better when the goods are not completely comparable, so I want to know if there are separability principles that yield plausible verdicts about cases involving parity. Here's the central puzzle case:

Two Blessed Lives

Tyche can bestow luck in love, and luck in work, and they are on a par. Today she will bestow luck on Jason and Jeff. Asclepius can bless the boys too, a small blessing that will add a month to each of their lives. The blessings of Tyche are much more beneficial though, and when enhanced by Asclepius' blessing, luck in love is still on a par with love in work (and vice versa).

Suppose Tyche blesses Jason with luck in love, and Jeff with luck in work. Will the outcome, which I'll call a *distribution*, be better if Asclepius adds his blessing to each?

Two Blessed Lives	Jason	Jeff
D1	Love	Work
D2	Love+	Work+
D3	Work+	Love+

The PPG says it will be, and that certainly seems to be right. Since D2 is better for each person, surely it would be pretty odd for someone to deny that it's Better. The PPG gets the right verdict and seems to give the right explanation for that verdict.

   Now suppose Asclepius perversely declines to add his blessing unless Tyche reverses hers. He will give each boy an extra month of life, but only if

Tyche bestows luck in work on Jason and luck in love on Jeff. (In the table, that distribution would be D3.) Work+ isn't better than Love, nor is Love+ better than Work, so D3 is *not* better for Jason or Jeff than D1. Given that the extra month counts as a small improvement (and if you think not, just shorten it), it cannot break parity. That means the PPG is silent. If we want a principle that tells us how D3 compares to D1, we need a louder principle. I will consider a natural one in Section 3, but first let me say more about what separability is and how it works.

2. Separability Principles

At least typically, the ordering of distributions from better to worse is a function of how good things are for the persons in the distribution. Distributions can be thought of as *vectors*, n-tuples whose locations are persons and whose elements are the goods those persons have in the distribution. For example, in the Two Blessed Lives table, each row is a distribution vector, an ordered pair of kinds of luck for Jason and for Jeff. The rows are vectors in the dimension of persons.

There can be vectors of goods also in other dimensions; in particular we'll be interested in the dimension of possible states. The dimension of time is also of some interest, but I won't be able to consider the aggregation of value over time in this chapter.[5] There are principles like PPG for each dimension, saying that the options are separable in that dimension.

There is a formal, mathematical definition of separability. It uses the idea of a utility function that *represents* an ordering. This just means that the function assigns numbers to items in the ordering in such a way that items higher up get larger numbers, so that x is better than y if and only if $u(x) > u(y)$, and $u(x) = u(y)$ if and only if x and y are equally good.[6] When an ordering is represented by a utility function, separability says that the ordering of the vectors is representable by a utility function that is an *increasing function* of utility functions that represent the ordering of the items in the locations on the vector. (This is *weak* separability. *Strong*

---

[5] See Broome (1995), and also Velleman (1991), for some reasons why separability in time is not particularly plausible.

[6] Standard utility functions have *real numbers* as values. Thanks to an anonymous referee for pointing out that Easwaran (2014) builds a framework for a more general conception of utility, in which *differences* between pairs of outcomes are fundamental and the values of the function can be comprised in more relaxed structures than the reals.

separability says we can carve out arbitrary subvectors, represent *their* ordering by utility functions, and then represent the ordering of the vectors with a utility function that is an increasing function of the utilities of the subvectors. The difference between strong and weak separability will not be important in this chapter.)

So that is the formal notion. It is of interest to philosophers because of what it says about how value aggregates. PPG is a principle of separability in persons, saying, in effect, that there isn't any value that isn't valuable for someone. Separability in the dimension of possible states, or ways that the world might turn out to be, is of great interest in decision theory. Indeed, it is one of the axioms of decision theory: the Sure-Thing Principle in the version in Savage (1972), Better Prizes in Von Neumann and Morgenstern (1944), and Impartiality in Jeffrey (1965). Decision theory is about preference, but the principle carries over naturally to axiology, since the 'better than' relation seems to work in a very similar way to the 'preferred to' relation. In this dimension, separability says that when one option is better in some state and at least as good in all states, it is better.

The axiological intuition behind Sure-Thing[7] is that wherever the value that makes one option better resides it must be in one of the possible outcome states. It cannot lie in the relation between possible outcomes, nor somewhere in the cracks between outcomes (if that makes sense). To see the force of this intuition, it helps to look at a potential counterexample. That will show what Sure-Thing rules out.

Suppose I have a kitten to give away. I can give it to Talia or to Horace. Talia is more excited about getting a kitten than Horace, but I think the kitten might be a little better off with Horace (and he does want the kitten!). All things considered, I think it is *slightly* better that the kitten end up with Horace.

It strikes me, though, that just giving one of my friends the kitten is a bit unfair. To be fair, I could flip a coin: if the coin lands tails I'll give the kitten to Talia, and if heads I'll give it to Horace. But Sure-Thing says the coin flip option has to be worse, since one of its outcomes is worse than simply giving the kitten to Horace, and the other outcome just *is* giving the kitten to

---

[7] I'll call the principle of separability in states "Sure-Thing", although Jeffrey's theory does not rely on the states of interest being probabilistically independent of which option is chosen, as Savage's Sure-Thing Principle does.

Horace. So if we think this is wrong, and flipping the coin is better, then we'll think Sure-Thing gives the wrong answer in this case.[8]

The point of introducing this example here isn't to refute Sure-Thing, but to illustrate how it is telling us that all value of a prospect is located in some outcome or other. Fairness, perhaps, is not located in any outcome in particular but in the relation between them: it's a matter of each friend's having some chance of getting the kitten.[9]

So much for separability in its usual guise: a principle for guiding and explaining value comparisons in cases in which orderings are complete. As we noticed when looking at Two Blessed Lives, the PPG is unhelpful in parity cases. Perhaps unsurprisingly, so is the Sure-Thing Principle, separability in states of nature. In the next two sections I will look at natural extensions of separability in these dimensions to see if they can help us in parity cases. To peek ahead: they do give us verdicts, but these verdicts are dubious and it looks like the beefed-up separability principles are not true. Then in Section 5 I will consider a different sort of principle for getting comparisons of vectors from comparisons of their elements, and comment on the interpretations of apparent incomparability that might motivate and rationalize this other sort of principle.

3. Separability for Parity in Distributions

We might try the following natural way of extending PPG so it tells us something about parity cases: the Negative Principle of Personal Good.

(NPPG)   A distribution cannot be Worse unless it is worse for someone.

This principle looks like it might be the same principle as the PPG, but it isn't. It will have something to say about parity cases, because it is about what is *not* worse. A key fact about parity is that it obtains when neither item is worse, and when neither item is better (and when they aren't equally good). Two Blessed Lives illustrates the extra content of NPPG. Since D1

---

[8] Diamond (1967), the source of this type of counterexample, suggests that Sure-Thing, or an equivalent separability principle, is okay as a constraint in decision theory on individual choice, but must be rejected for social choice, since social choice is more intrinsically concerned with fairness.

[9] This method of accounting for the value of fairness is not forced. I defend an alternative in Dreier (1996).

isn't worse for either boy than D3, the NPPG says, it isn't Worse than D3: it's on a par. Might D1 be exactly as good as D3? That verdict is consistent with the NPPG, but it looks like it couldn't be right. For compare D4: Work+ for Jason and unenhanced Love for Jeff.

Two Blessed Lives	Jason	Jeff
D1	Love	Work
D2	Love+	Work+
D3	Work+	Love+
D4	Work+	Love

The PPG says D4 is worse than D3. But the NPPG says it isn't worse than D1. So D1 and D3 can't be equally good. (Since D4 is worse than D3, it is worse than everything that's exactly as good as D3.) The NPPG, then, tells us that D1 and D3 are on a par, so it does give a verdict, unlike PPG.

NPPG looks like a separability principle too, but since we want to apply it in cases involving parity, for our purposes it cannot be formal separability. If there is such a thing as parity, then good cannot be represented by utility functions at all. For when one outcome or prospect is represented by a utility number, x, and another is represented by y, then, since x must be greater than, equal to, or less than y, the first outcome must be better than, exactly as good as, or worse than the second. (That is what it means, remember, for a utility function to *represent* an ordering; although see Note 6.) There is no parity among the real numbers, so there couldn't be parity among the items they represent as utility numbers. Still, NPPG captures the philosophical idea behind the formal notion of separability. All of the worse-making, it tells us, is worse-for-somebody-making. For example, some kinds of egalitarianism will be incompatible with the NPPG, since they will say that a less equal distribution can be worse just in virtue of its inequality, and not because it is worse *for* someone. And these forms of egalitarianism will also be incompatible with the PPG.[10] Separability principles rule out the possibility that there is value (or disvalue) in the cracks between persons, or fundamentally residing in the relations between them.

So NPPG is a good candidate for extending PPG to cases of parity. The problem is that there is a compelling reason to doubt that it is true.

[10] Broome (1995) calls them 'non-additively-separable communal egalitarianisms'. Ch. 9 provides a helpful discussion.

The Problem with NPPG

The NPPG conflicts with

(Neutrality)   When one distribution is a permutation of another, they are equally Good.

A distribution that is a permutation of another distribution differs from it only in that it assigns the goods to different people. Neutrality, then, says that the determination of which distribution is better has to be insensitive to the identities of the bearers of the goods.[11] There are many sorts of cases in which Neutrality is wrong. Suppose the goods are hand tools. Then if Jeff is highly skilled with a jigsaw and Jason is hopelessly inept with the jigsaw but swings a mean hammer, distributing the jigsaw to Jeff and the hammer to Jason is better than the converse. A similar caveat applies to the example we're interested in. Not everyone benefits as much from luck in love as they benefit from luck in work; presumably one kind of luck is a lot more important for some people, and the other kind is a lot more important for others. We are assuming, in our example, that either Jason and Jeff are similar enough, or else that the manifestations of luck will be shaped and tailored to the contours of its recipient's needs and idiosyncrasies; we are assuming that each boy's benefit from a kind of luck will be the same as the other boy's. I'll return to this complication, and provide a reformulated, hedged version of Neutrality, in the next section.

So here's the powerful Neutrality Argument: Neutrality tells us that D3 and D2 are equally good. But PPG tells us that D2 is better than D1. So, D3 is better than D1, *pace* NPPG.

As Nebel (2020) notes, this problem with NPPG (which he calls the *person-affecting restriction*) is closely related to Parfit's (1986) Non-identity Problem. Suppose that if they conceive now a certain couple will have a child with a disability that will make the child's life worse (than if she didn't have the disability). If they delay two months, they will conceive a different child who will not suffer from the disability. Suppose further that (implausibly!) the couple's own lives will not go better or worse if they delay than if they

---

[11] Neutrality for distributions is what is called 'Anonymity' in social choice theory, except there the vectors have welfare utility numbers as their elements, rather than actual goods. See, for example, List (2013).

conceive now. Call the two possible children 'Earl' and 'Leia'. Then the distributions the couple choose between are:

NonID	Earl	Leia
Now	Life	____
Delay	____	Life+

In the first distribution, Leia does not exist, so she has no entry in the table; likewise for Earl in the second distribution. So, in neither case is there anyone for whom either distribution is better or worse, so the NPPG says neither can be better or worse than the other. Parfit (and many others) think that it is better to Delay conception even so. Nebel's point is that the problem for the NPPG can arise even when the same persons exist in both distributions, if there are possibilities of lives on a par.[12] I think there is more to be said: nonexistence, I think, might best be seen as on a par with existence.

In Broome (2004), John Broome sets up (and then tackles) the Non-identity Problem by putting 'Ω' into tables as a placeholder for the good for a person in a distribution in which that person does not exist. If the value of nonexistence could be given a number, equal to the value of some particular sort of life, say, then we could think of Ω as a constant. But Broome, and others, think that adding new people to a population is very typically *neutral*, making the new distribution neither better nor worse, and this is so for a broad range of different sorts of lives the new people might have. For example, take the distributions represented in the Populations table (below). The columns represent groups of persons. Small is a small population; the other two add new people, either not as well-off as the original ones or exactly as well-off as them. Broome thinks that the Unequal distribution in the table is no better or worse than the Small one, and the Equal distribution is also no better or worse than the Small one. It can't be that both the Equal and the Unequal distributions are exactly as good as the Small distribution, since (by transitivity of *exactly as good as*) they would then be exactly as good as one another, which they are plainly not. So, Small is on a par with the other distributions, and so it looks like Ω is on a par with levels of well-being represented by the different numbers.

Populations	G1	G2
Small	4	Ω
Unequal	4	3
Equal	4	4

So NPPG, the natural parity extension of PPG, looks unreliable. The argument against it appeals to two substantive principles: Neutrality and the PPG. These two principles are not self-evident truths. It is hard for me to believe that they are wrong in this case, but there might be room for a powerful and otherwise plausible theory to deny one of them. The argument against NPPG also appeals to transitivity, but that is a *logical* principle. It is built in to the *Better than* relation, because 'Better' is a comparative.[13]

4. Separability for Parity in Prospects

Miriam Schoenfield has offered an extension of separability for prospects in cases involving parity.

> LINK: In cases in which considerations of value are the only ones that are relevant, if you are rationally certain that one option, A, will bring about greater value than the alternative option, B, you're required to choose A. If you are rationally certain that neither of the two options will bring about greater value than the other, it's not required that you choose A, and it's not required that you choose B.   (Schoenfield 2014)

LINK says, in effect, about prospects what NPPG says about distributions. It is a kind of separability for parity, although again it cannot be formal separability. The idea that a prospect cannot be worse unless it is worse *in some possible state* can generate its own Nonexistence Problem, just in the way that the idea that a distribution cannot be worse unless it is worse for someone generates the Non-identity Problem, though as far as I know this state version is so far undiscussed. Suppose Tyche, now in her more chaotic mode, decides to bring someone into existence, say John, if her coin of fortune lands heads, but not if it lands tails. Asclepius offers to bless John

---

[13] In case this is not convincing just by itself, see Hedden (2020) for a thorough defense.

with immunity to the common cold . . . but only if Tyche will reverse her coin plan and bring John into existence if and only if it lands tails.

NonID	Heads	Tails
No blessing	Life	Ω
Blessing	Ω	Life+

Maybe those who find it obvious that the couple should delay the conception of their child to avoid the birth defect will also think it's clear that it is better if Tyche accepts Asclepius' offer. But this is not what LINK says. Tyche can be rationally certain that things won't be worse if she declines the offer—if the coin lands heads, there will be someone whose life is on a par with non-existence, and if it lands tails, there is nobody at all, on a par with the life that would have existed had she taken Asclepius' offer.

If this is right, then the problem exposed in Parfit's Non-identity example does not depend on the two possible lives being attached to two distinct persons. It is enough that nonexistence is on a par with a range of possible lives.

Weak LINK?

In Section 2 we saw that there is a powerful reason to think that NPPG isn't true. And LINK is a closely-related separability principle. Here is an analogous argument against LINK.

Lucky You

Suppose instead of distributing luck to two different people, Tyche flipped a coin to decide which kind of luck to give you. She will give you luck in love if the coin lands Heads, and luck in work if it's Tails. Asclepius again perversely offers to add his blessing only if Tyche will reassign her blessings to the outcome of the coin toss, associating luck in love with tails and luck in work with heads.

Lucky You	Heads	Tails
P1	Love	Work
P2	Love+	Work+
P3	Work+	Love+

Now the rows in the table are what I called prospects, not distributions. Distributions, remember, are outcomes in which the goods are distributed among persons. Prospects are items that we can choose or evaluate, in which the goods are distributed though the space of possibilities, for example when there will be different outcomes depending on whether the coin lands heads or tails. Prospects are like distributions in the dimension of possible states.

LINK operates the way NPPG did, to tell us that it is okay for Tyche to go with the first prospect, declining Asclepius' offer. For in neither state of the coin does P3 bring about greater value than P1. In each state, the outcome of P3 is on a par with the outcome of P1.

LINK does not tell us which prospects are Better, but only which we are permitted to choose, so we cannot refute it with a Neutrality principle.[14] But, if we can rely on this other Neutrality principle—

(Prospect Neutrality)   When one prospect is a permutation of another, they are equally good.

—we can conclude that P3 is in fact worse than P1. For P2 is exactly as good, for you, as P3, according to Prospect Neutrality. And it is better for you than P1, so P3 must be better (for you) than P1. And if it's better for you, then it is hard to believe that when only considerations of value are in play it is permissible to decline it and choose P1 instead. So that is bad news for LINK.

In one way, this argument is weaker than the argument against the NPPG, and in another way it is stronger. It is weaker in that it is not completely clear how to decide when a *prospect* is better or worse for a person than another prospect. But it is stronger in that the Neutrality principle employed against the Negative Better Prospects Principle is on more solid ground than the one employed against the NPPG.

The Weakness in the Argument against LINK

Let's start with the weakness, which in my opinion is not very serious.

Some people think that since we don't know how prospects will turn out, we don't know (usually) which ones are better for someone than which; they think, in fact, that prospects themselves are not better or worse, for a person,

---

[14] Doody (2019) argues against what he calls the *Never Worse* principle, but despite its name it is also about preferences.

since only their outcomes are better and worse. So that would mean the concepts we're using in the argument against LINK are dubious. But I think this worry is misguided.

First of all, it is not common sense. Surely common sense says that it is better to get a ticket in a lottery for a trip to Paris than it is to get a ticket in a lottery for a trip to Peoria. (Of course, assuming the player has common tastes and preferences, and that the two lotteries give equal chances at winning.) And if it's not better *for you*, then how is it better? In the second place, in general (though not always) the outcomes in a prospect are themselves evaluable only as prospects. In common examples the outcomes of the prospects are money prizes. In the ones I'm using, the outcomes are trips to interesting or fun places. But getting, say, a thousand dollars is only *probably* good, and how good it is plainly depends on some uncertain future possibilities. For instance, if we are heading into a period of rampant inflation, the money prize is less valuable than if we are in a deflationary economic period. And how good a trip to Paris is depends on whether Europe will be in the midst of a pandemic, among other things. If these 'outcomes' can be evaluated as better and worse, then *of course* prospects can be, for the outcomes *are* prospects.

The Strength in the Argument against LINK

So I think the objection that prospects themselves can't be better or worse for you is not a good objection. Of course, it can be hard to decide which prospects are better. But in the cases of interest to us, it doesn't seem hard at all. The prospect in which both outcomes are enhanced by Asclepius' blessing is surely better than the unenhanced ones, since it is better in every possible outcome state. And the prospects P2 and P3 are surely equally good, since heads and tails are equally likely. In this judgment we rely on a *much* weaker principle (namely, Prospect Neutrality) than decision theorists rely on. Neutrality is acceptable to Risk-Weighted Expected Utility maximizers, for example (indeed, it is not just acceptable but follows from their view).[15] Suppose someone thought P2 was better than P3. We would conclude, I think, that she thought the coin was biased, or that it was important whether the coin landed heads or tails, or that somehow the

---

[15] It is even weaker than Stochastic Dominance, which REU Theory proudly accepts as a replacement for Sure-Thing; see Buchak (2017).

kinds of luck interacted with the face of the coin (so that luck in love has a weaker effect when coins in the neighborhood land heads). Implausible as this is, it's the only way *I* can think of to make sense of the view that one of those two prospects is better. And this is the *strength* of the argument that relies on Neutrality for prospects, by comparison with the one that relies on Neutrality for distributions.

It is not as obvious that permuting benefits in the dimension of persons is value-neutral as it is that permuting them through equiprobable states is value-neutral. There are some similarities—I mean, there are some ways in which Neutrality could fail in the dimension of persons that would be analogous to a way it could fail in the dimension of states. I am going to spell these out. The ways in which Neutrality could fail illustrate the content of the principle. After I give the examples, I'll hedge the Neutrality principles against these sorts of failures. Then we'll have weaker principles, but they will still have something to say about the cases of interest to us. My hypothesis is that Neutrality in the dimension of persons will turn out to be less compelling than Neutrality in the dimension of states. And that in turn will be a strength in the argument against LINK, as compared with the argument against NPPG.

Neutrality Failures and Hedging

First, here is a very plain way Neutrality in persons could fail: if the distributions had groups as their locations instead of individual persons, it could be better to distribute the benefit of Good Weather to a very large population, rather than a small population; similarly it is better to have the greater benefit in the more probable state of nature. In many plausible axiologies, population of a group and probability of a state determine the weightings of goods in those locations. But in the cases that we're really interested in, the locations on our distributions are individual persons, not groups. Still, maybe there are cases in which one person should be given more weight than another?

I think the most threatening example involves bestowing benefits (or costs) on persons, some of whom are much worse off than others. Suppose the benefit is a full day of pain relief. Each potential beneficiary, Mack and Milly, has a migraine today. The distribution will either consist of Mack having a headache today and Milly being pain-free, or vice versa. So the distributions are permutations of one another. But maybe the distributions are not equally Good because Mack suffers from regular migraines, so he is worse off than Milly. Maybe it is Better to give the benefit to the less well-off person.

Well, for one thing, we *might* think that really this is because a pain-free day is more important to Mack than it is to Millie, so the benefit is greater. Plausible, but it's also reasonable to think that even if it isn't better *for Mack*, it's still Better to give him the benefit, perhaps because of the Difference Principle, or because Prioritarianism is true.[16] So that's the first threat to Neutrality: unequal weighting.

Second, as I mentioned in Section 3 when I first discussed Neutrality, sometimes a given benefit might be much more important for one person than for another. The gift of fleetness, bestowed by Artemis, was of much greater value to Atalanta than it would have been to Archimedes. This violation of Neutrality has to do with the *interaction* between the location, in this case a person, and the type of benefit. And similarly, sometimes an outcome interacts with a state of nature: a prospect offering a trip to Truro Beach if it is raining and to the Boston Museum of Science if it is sunny is not as good at the permutation of that prospect, even if the types of weather are equally likely.

But neither of these sorts of problems for Neutrality—extra weighting, interaction between location and outcome—is relevant in our examples. So we can hedge our Neutrality principles, and they'll still apply to the cases that interest us. Distributions and prospects are each vectors; for distributions the locations on the vectors are persons, and for prospects the locations are states. So the general form of Neutrality, now suitably hedged, is this:

(Hedged Neutrality)   When each location on a vector gets equal weight, and there is no value interaction between locations and located goods, a vector and its permutation are equally good.

This Hedged Neutrality principle is not stated in a very rigorous way, since I have left the idea of 'interaction' rather loose and intuitive. But I hope it is

---

[16] For Prioritarianism, see Parfit (1997); for the Difference Principle, of course, Rawls (1999). It is a somewhat thorny question whether the idea of priority also implies that some *states*, namely those in which a person is badly off, should be given greater weight than others (in which the person is well-off). I think in the end the answer is *no*. Whether or not the idea that benefits to badly off persons are *morally* more important is true, it certainly makes sense. Whereas I doubt that the idea that increases to your welfare in states in which you are badly off are more important than increases to your welfare in states in which you are well-off, where the 'more important' means more beneficial to *you*, is really a coherent idea.

clear enough. In the examples we are considering, there doesn't seem to be, intuitively speaking, any interaction between states or persons and their located goods. So the 'hedged' version of Neutrality will be plausibly true, and it will still apply to our examples.

What seems less compelling, to my mind, about Hedged Neutrality for distributions is that it is not obvious that *the identity of persons* is itself irrelevant to how valuable it is that this or that person receive a benefit. By contrast, the mere identity of states could not possibly matter. To see this, maybe it's enough to observe that in lots of presentations of decision theory we don't even label the columns of our table with state descriptions; we just write in probability numbers! When we do, it would be extremely peculiar to resist the Neutrality principle when equiprobable columns have been permuted. We'd think the resistor must be misunderstanding our representation. Now, in the examples I've given, the states are coin flip outcomes. But it's a kind of trope or convention that these are simply stand-ins for any old equiprobable states that don't matter at all (to the agent's preferences, in decision theory; to what's good or bad, in axiology). They don't matter in themselves, and they don't interact significantly with the outcomes in their locations.[17] (It would be a violation of the convention to use a coin flip to determine whether you got as a prize a ticket worth $1 if the next coin flipped lands heads or a ticket worth $100 if the next coin flipped lands tails.)

I do not know of an example in which the distinctness of persons seems to matter in a way that would threaten Neutrality. But the fundamental idea of Neutrality does not seem as obvious in persons as it does in states. It seems obvious that when neither state gets more weight (by being more probable) and there is no interaction between a state and the good occurring in that state, swapping goods between state locations must leave a prospect exactly as good as before. It does not seem so obvious that it couldn't matter which person gets some good, even if they will benefit equally and neither person is more important.

The Neutrality argument against NPPG is not conclusive. Against the extension of Sure-Thing for incomplete Betterness it is more powerful, and suggests that LINK is mistaken. It's worth looking for an alternative to the separability approach to aggregation and comparison in cases of parity.

---

[17] As I interpret him, a proposition picking out a state that "doesn't matter" in this way is what Ramsey (1990) means by an "ethically neutral proposition", although other commenters take him to mean more simply just a proposition whose truth the agent doesn't care about, without reference to interaction with the goods in the outcomes.

5. An Alternative to Separability, and Other Interpretations

Caspar Hare offers such an alternative. His interest is in incomplete preferences, not axiological parity, but an analogous approach could help us. Hare's *Prospectism* says:

> It is permissible for me to choose an option iff, for some utility function that represents a coherent completion of my preferences, *u*, no alternative has greater expected *u*-utility.   (Hare 2010)

Remember: when an ordering is incomplete, it cannot be represented by a utility function, since utility functions have values in the real numbers, and the real numbers are strictly ordered. But, an incomplete ordering can be represented by a *set* of utility functions, each 'consistent with' the incomplete ordering. That is, for each utility function *u* in the set, and for every pair of items $<x, y>$ that *are* ordered, $u(x) > u(y)$ iff *x* is ranked together with or above *y*. Each of these utility functions 'represents a coherent completion', as Hare says. The intuition behind Prospectism, then, is that if *some* utility assignment that could complete the incomplete ordering also gives a given prospect at least as great a utility than a given alternative, it should be okay to pick that prospect rather than the alternative. But if *every* utility assignment that could complete the incomplete ordering gives this certain prospect a lower ranking than an available alternative, you must not pick that prospect over alternatives.

Let's replace talk of utility with the underlying ordering, and apply the general idea to axiology. By a 'coherent completion', Hare means a completion that obeys the Sure-Thing Principle.[18] So our version would be:

If a prospect P is at least as good as a prospect Q in every completion of the ordering that obeys the Sure-Thing Principle, then P is at least as good as Q.

And we can add,

If P is at least as good as Q in *some* completion of the ordering that obeys the Sure-Thing Principle, then Q is not better than P.

---

[18] At least I think this is what he means. When he applies his principle, he seems to be assuming that all the utility functions are *expectational*; that can be true only if the orderings they represent are separable in states of nature.

A similar approach could be applied to distributions and complete orderings that obey the PPG. In this chapter I'll discuss only the extension to prospects, leaving any special features of distributions to another occasion.

The first part of this principle tells us that P3 is better than P1. For a utility function will have to give Work+ a higher number than Work, and Love+ a higher number than Love, so the expected value of the prospect <Work+, Love+> has to be greater than the expected value of <Love, Work>. (The coin is fair, so the expected value is just half the sum of the utilities in the prospect.) Prospectism is consistent with Neutrality. (And the parallel principle, which unfortunately would have the ugly name 'Distributionism', is consistent with Neutrality for Persons.)

Doody (2019) takes Prospectism to be more promising than LINK, largely because it is compatible with a form of dominance that is weaker than Sure-Thing. But there is something odd about using Prospectism to handle cases involving *parity*.

To see why, let's first think about some other interpretations of the Apparent Incomparability phenomenon. One interpretation is epistemic: the reason we have trouble placing apparently incomparable items with respect to one another is due entirely to our ignorance. For one thing, we haven't been given much detail about what luck in love and luck in work amount to! And we probably don't know enough about the trips to know which is better for Mark. And even if we were very knowledgeable about the details of the trips or luck, it might be very difficult to know which kind of luck would work out for a specific person, or which trip would be better for Mark in particular. And then even if we knew a huge amount about the person, it could be that there are permanent obstacles to our full understanding of exactly how non-ethical features of a person's life contribute to the person's welfare. Maybe there is a true and complete theory of welfare, and if we knew it and had enough information about Jason and Jeff's lives and Mark's interests, what looked like incomparability would be revealed as mere uncertainty. The first half of Prospectism looks very sensible, under this interpretation, and the second half is also fine if instead of concluding that P2 is not better than P1 we conclude that it isn't *knowably* better.

A second interpretation is the Indeterminacy interpretation. According to this conception of apparent incomparability, there is just no fact of the matter as to which life is better. Maybe this is because our concepts of *better* and *worse* have fuzzy edges, or were otherwise not constructed for such precise comparisons; think of the various reasons philosophers have for thinking that certain of our predicates are vague. Or maybe it's a more metaphysical indeterminacy, as perhaps it is indeterminate whether an atom of francium would have decayed by now if one had been created twenty minutes ago. To be clear: if that is the correct interpretation of our inability to connect two blessed lives by *better, worse, exactly as good*, it is *not* a case of parity. When two lives are on a par, it is not indeterminate whether the one with luck in love is better. When they are on a par, it is completely determinate that neither is better.

Prospectism looks pretty good on this interpretation, too. When a comparison is indeterminate, our concepts or the world just haven't settled which way it goes. But suppose one prospect turns out to be better than a second no matter *which* way the unsettled comparison goes. Then surely it's just better. Suppose in our comparisons, for example, we just haven't decided yet whether a certain small increase in the longevity of a pleasant experience outweighs a greater intensity in an experience we're comparing it with, but whichever weighting we give, a prospect giving some chance of each will be better in each state than an alternative prospect. Surely the prospect offering better outcomes no matter how we resolve the indeterminacy is just better.

On either of these interpretations, LINK doesn't really apply. It only issues a judgment when a prospect is not better than a second in any state. But on the indeterminacy interpretation, it's indeterminate whether the prospects are better, so the condition isn't met (or maybe, it isn't determinately met!), and on the epistemic interpretation, we just don't *know* of either state that the prospect is better in that state. What if we construct a version of LINK for use in cases of indeterminacy, and one for ignorance?

INDETERMINATE LINK   If a prospect P is not determinately better than a prospect Q in any state, then it is not determinately better.

IGNORANT LINK   If for each state we can't tell whether P is better than Q in that state, then we can't tell whether P is better than Q.

But these are dubious principles, basically for the same reason that Prospectism is plausible on these interpretations. Even if we're not sure whether the white bean soup is better than the kale salad, and we still can't tell even when the white bean soup gets some grated Parmesan, we could be quite sure that the kale salad plus grated cheese *and* the white bean soup with grated cheese are together better than the unenhanced appetizers together. Or suppose we compare the length in words of *Moby Dick*, with one page of the long description of blubber-rendering removed, with *Harry Potter and the Deathly Hallows;* and then we compare *Harry Potter and the Deathly Hallows* with a page of relationship difficulties between Hermione and Ron edited out with unabridged *Moby Dick*. I can't tell which is longer, in either case! But I know for sure that the two unabridged tomes together are longer than the two abridged ones. And, I think, the situation is similar when the phenomenon is interpreted as indeterminacy. If we simply asked about the *length*, without specifying whether it was to be calculated in words or pages, that could leave each initial comparison vague and indeterminate, even though the comparison of the pairs is perfectly determinate.

So Prospectism looks like it makes good sense and is more successful for use when our orderings are incomplete because of indeterminacy or ignorance. But, as I said, there is something odd about using Prospectism to handle cases of parity. Its reasoning seems to go like this: if no matter how you complete this incomplete ordering, *x* turns out better than *y*, then we can conclude *x* just *is* better. And although this seems to make some sense when we regard the incompleteness of our ordering as, say, inability to make up our minds, or the result of an unfinished concept—something whose nature suggests that it is to-be-completed in some way or another, but it's just not decided which way—that is not how parity works. When luck in love and luck in work are on a par, that's not because we haven't finished the job of deciding or sanding down the edges of our concept. It's because neither of them is at least as good as the other—they just don't stand in any of the more common comparative value relations, as Chang (2002) explains. To say, "Right, but no matter how we got rid of parity in favor of the usual relations, this prospect would turn out to be better than that prospect, so really it must just *be* better," just seems to be a non sequitur. It's as though, having heard that the number six has no spatial location at all, someone replied, "Right, but no matter how we assigned it a spatial location, it would be reachable from Tucson in some finite amount of time by a light wave, so we can conclude that it just *is* reachable from Tucson in a finite amount of time." This is (very) bad reasoning precisely because having no spatial location isn't the same thing as having some indeterminate spatial location, or having an unknown and unknowable spatial location.

Conclusion

I wanted to find principles that could help make and explain comparisons between options in cases of apparent incomparability in the way the Principle of Personal Good can explain them when the ordering is complete, namely, in terms of the comparisons of their parts. The negative separability principles do not fare well. Some hope can be found in the fact that the Neutrality argument against the NPPG is not a knock-down argument, since the Neutrality principle is not itself obvious, and seems more doubtful than its counterpart LINK.

Supervaluation-inspired principles, like Hare's Prospectism, are consistent with Neutrality and yield better answers. Prospectism makes good sense if incomparability is a matter of ignorance or indeterminacy, but not, I argued, when it is a matter of parity. An explanation of comparisons of vectors (whether they be prospects or distributions) in terms of comparisons of their parts in cases of parity is elusive.

References

Broome, John. 1995. *Weighing Goods: Equality, Uncertainty and Time*. Oxford: Wiley-Blackwell.

Broome, John. 2004. *Weighing Lives*. 1st edn. Oxford: Clarendon Press.

Buchak, Lara. 2017. *Risk and Rationality*. New York: Oxford University Press.

Chang, Ruth. 2002. "The Possibility of Parity." *Ethics* 112 (4): 659–88. https://doi.org/10.1086/339673.

Diamond, Peter A. 1967. "Cardinal Welfare, Individualistic Ethics, and Interpersonal Comparison of Utility: Comment." *Journal of Political Economy* 75 (5): 765–6.

Doody, Ryan. 2019. "Opaque Sweetening and Transitivity." *Australasian Journal of Philosophy* 97 (3): 559–71. https://doi.org/10.1080/00048402.2018.1520269.

Dreier, James. 1996. "Rational Preference: Decision Theory as a Theory of Practical Rationality." *Theory and Decision* 40 (3): 249–76. https://doi.org/10.1007/BF00134210.

Easwaran, Kenny. 2014. "Decision Theory Without Representation Theorems." *Philosophers' Imprint* 14.

Geach, P. T. 1956. "Good and Evil." *Analysis* 17 (2): 33–42. https://doi.org/10.2307/3326442.

Hare, Caspar. 2010. "Take the Sugar." *Analysis* 70 (2): 237–47. https://doi.org/ 10.1093/analys/anp174.

Hedden, Brian. 2020. "Consequentialism and Collective Action." *Ethics* 130 (4): 530–54. https://doi.org/10.1086/708535.

Jeffrey, Richard C. 1965. *The Logic of Decision.* Chicago: University of Chicago Press.

List, Christian. 2013. "Social Choice Theory." In *The Stanford Encyclopedia of Philosophy,* ed. Edward N. Zalta, Winter 2013. Metaphysics Research Lab, Stanford University. https://plato.stanford.edu/archives/win2013/entries/ social-choice/.

Nebel, Jacob M. 2020. "A Fixed-Population Problem for the Person-Affecting Restriction." *Philosophical Studies* 177 (9): 2779–87. https://doi.org/10.1007/ s11098-019-01338-5.

Parfit, Derek. 1986. *Reasons and Persons.* Oxford: Oxford University Press.

Parfit, Derek. 1997. "Equality and Priority." *Ratio* 10 (3): 202–21. https://doi.org/ 10.1111/1467-9329.00041.

Ramsey, Frank Plumpton. 1990. "Truth and Probability." In *Philosophical Papers,* ed. D. H. Mellor, 52–94. Cambridge: Cambridge University Press.

Rawls, John. 1999. *A Theory of Justice.* 2nd edn. Cambridge, Mass.: Belknap Press.

Savage, Leonard. 1972. *The Foundations of Statistics.* New York: Dover Publications.

Schoenfield, Miriam. 2014. "Decision Making In the Face of Parity." *Philosophical Perspectives* 28: 263–77. https://doi.org/10.2307/26614553.

Thomson, Judith Jarvis. 1997. "The Right and the Good." *Journal of Philosophy* 94 (6): 273–98. https://doi.org/10.2307/2564542.

Velleman, J. David. 1991. "Well-Being and Time." *Pacific Philosophical Quarterly* 72 (1): 48–77. https://doi.org/10.1111/j.1468-0114.1991.tb00410.x.

Von Neumann, John, and Oskar Morgenstern. 1944. *Theory of Games and Economic Behavior.* Princeton: Princeton University Press.

6

The Reasons Aggregation Theorem

Ralph Wedgwood

0. Aggregating Reasons

In many cases, when an agent faces a choice between alternative courses of action, there are *reasons* both *for* and *against* each of these alternatives.

For example, suppose that you face a choice between two Italian restaurants, Luna Caprese and Sole Siciliano:

	Food quality	Economy
Luna Caprese	Reason *For*	Reason *Against*
Sole Siciliano	Reason *Against*	Reason *For*

Suppose that considerations of food quality tell in favour of Luna Caprese and against Sole Siciliano, while considerations of price and economy tell in favour of Sole Siciliano and against Luna Caprese. This is a situation in which there is *one* reason—grounded in considerations of food quality—in favour of Luna Caprese and against Sole Siciliano, and also *another* reason—grounded in considerations of price and economy—in favour of Sole Siciliano and against Luna Caprese.

Nonetheless, it might be clear that *overall*, or *all things considered* (ATC), there is *more reason* for one alternative than for another. For example, perhaps Luna Caprese is only *slightly* better than Sole Siciliano in terms of food quality, while Sole Siciliano is *much* better than Luna Caprese in terms of economy. Then overall, or ATC, there is more reason to choose Sole Siciliano than Luna Caprese.

Somehow, then, there must be a way of *combining* or *aggregating* the different reasons for and against each of the relevant alternatives, to yield an overall or ATC verdict on *how much reason* ATC there is for the agent to take each of these alternatives.

Ralph Wedgwood, *The Reasons Aggregation Theorem* In: *Oxford Studies in Normative Ethics, Volume 12*.
Edited by: Mark Timmons, Oxford University Press. © Ralph Wedgwood 2022.
DOI: 10.1093/oso/9780192868886.003.0007

This overall or ATC verdict on these alternatives is a conclusion about *how much reason* ATC there is for the agent to take each of these alternatives. There is a way of understanding 'should' and 'ought' so that if, out of *all* available alternatives, one has most reason ATC to do *A*, then *A* is what one ATC *should* or *ought to* do. In this way, the correct way of aggregating the reasons for and against the available alternatives determines what one (in this sense) ATC should or ought to do.

How can these different reasons be aggregated? On many ways of thinking about reasons, this is a deep and inscrutable question—perhaps even a question where no systematic answers are to be found. In this chapter, I shall sketch an alternative view of reasons, on which this question has a quite straightforward answer.

In this chapter, I shall sketch a number of premises about reasons, and about how they can be aggregated. Some considerations will be given in support of each of these premises, but I will not be able to offer anything like a full defence of these premises here. The main goal of this chapter is to show that, once all of these premises are in place, we can derive an analogue of a certain famous theorem from social choice theory. According to this theorem, the aggregation of reasons is necessarily *additive*. After explaining the theorem, I shall briefly consider some objections to this kind of additive aggregation. As I shall argue, these objections can all be addressed. In this way, I hope to marshal some considerations in favour of this essentially additive conception of the aggregation of reasons.

1. Reasons for Action Are Grounded in Values

Many philosophers hold that there is an intimate connection between *reasons for action* and *values*—where every "value" is a *way* of being *good*. As Joseph Raz (1999: 23) puts it: "reasons [for actions] are facts in virtue of which those actions are good in some respect and to some degree".

For example, the fact that I promised my mother to put up the Christmas decorations is a reason for me to put up the Christmas decorations because it is a fact in virtue of which the action of my putting up the Christmas decorations is *good*. Specifically, it is a fact in virtue of which the action of my putting up the Christmas decorations is an act of *fidelity* or *promise-keeping*—which is one respect in which an action can be good to a non-trivial degree.

If this is true, it will surely also be true that a reason *against* an action is a fact in virtue of which that action is *bad* in some respect and to some degree.

The fact that your pushing a certain button would be an act of torturing another person is a reason *against* your pushing the button, because it is a fact in virtue of which your pushing the button is a *bad* thing to do. Specifically, it is a fact in virtue of which your pushing the button is an act of *victimizing* and *exploiting* the other person, or *infringing that person's rights*—which are respects in which actions can be bad to a non-trivial degree.

As I shall put it here, every reason for an action is "*grounded*" in a certain way or respect in which the action is good. Each of these "ways" or "respects" in which actions can be good is what I shall call a "value". I shall assume that there are many different values of this kind—that is, there are many different ways in which actions can be good or bad. For example, there is the value of promoting self-interest, and there are also numerous moral values as well—such as fidelity, fairness, beneficence, non-maleficence, and so on. These values can *conflict* with each other—as when one action A is better than an alternative action B in terms of self-interest, even though B is better than A in terms of one of the moral values, such as fidelity or fairness or beneficence.

This, then, is the first premise of my derivation of how the aggregation of reasons is possible—the premise that all reasons for action are grounded in values in this way.

This premise is compatible with many different views of how reasons are grounded in values. For example, it is compatible with the consequentialist view that all reasons for actions are grounded in the values that are *promoted* by those actions, and also with the non-consequentialist view that some reasons for action are grounded in the values that are instantiated by the *actions themselves*.[1]

This premise is also compatible with a *desire-based* view of reasons (similar to that of Schroeder 2007). This is because one way in which something can be good is by being *good for* achieving an end E—where E is an end that the agent desires. Finlay (2014) argues that *every* normative use of 'good' expresses an end-relational way of being good of this kind. Even if this view is mistaken (as I believe), we may assume that the values—the ways of being good—include such end-relational ways of being good. (This assumption clearly guarantees the existence of a huge number of values—since it implies that for every end E, one of the values that exists is the value of being good for achieving E.)

---

[1] For a sketch of this form of non-consequentialism, see Wedgwood (2009).

Different theories of reasons will give different accounts of *which* of the many different values that exist are the *reason-providing values*, for the situation of an agent at a particular time. For example, the desire-based view of reasons will make the following two claims: first, whenever the agent desires an end E at a particular time, the value of being good for achieving E is a reason-providing value for that agent at that time; and secondly, there are no other reason-providing values.

I am in fact myself sympathetic to a much more *objectivist* view of the values that ground reasons for action. According to this more objectivist view, all the reason-providing values are more "objective" values that can be exemplified by actions—such as the value of self-interest, and the various moral values (such as fidelity, fairness, and beneficence) that I have canvassed above. However, for the purposes of this chapter, I can remain neutral about which values are the reason-providing values for an agent at a time. The first premise of the derivation that I shall discuss this chapter is simply that all reasons for action are grounded in values of some kind.

2. Reducing Goodness to Betterness

The second premise that I shall discuss here is a version of John Broome's principle that "goodness is reducible to betterness". Broome (1999: 163) formulates this principle as follows:

> If you knew everything about betterness—of every pair of alternatives, you knew whether one was better than the other, and which—then you would know everything there is to know about goodness. There is nothing more to goodness than betterness.

Specifically, as Broome (1999: 166) claims, "'X is good' reduces to 'X is better than the standard'"—where this standard of comparison is simply determined by the context in which the term 'good' is used. In the case of *actions*, the standard of comparison seems typically to be the contextually salient alternative actions. In short, to say that an action is good is simply to say that it is better than the salient alternative(s).

To be precise, I should note that my version of this premise is slightly different from Broome's. First, as I have explained in the previous section, I am assuming here that there are many different values—many different ways in which things can be good, or many different kinds of goodness. So, the premise that I need for my argument is that *each* of these kinds of goodness is reducible to the *corresponding* kind of betterness.

Moreover, I shall also allow that one may need to know a bit more about a given kind of betterness than the facts that Broome explicitly mentions, in order to know "everything that there is to know" about the corresponding kind of goodness. Specifically, one may also need to know, not only whether one of the two alternatives in question is better, and if so which, but also *how much* better (in terms of the relevant value) the better of the two alternatives is. For some purposes, one may also need to know the *explanation* of all these facts: that is, one may need to know *why* A is better than B, to the extent that it is, or why C is not better than D, and so on.

Apart from these qualifications, however, the second premise of the derivation that will be discussed here is fundamentally similar to Broome's principle that "goodness is reducible to betterness". Combining Raz's view of reasons with Broome's view of goodness yields the following two principles:

- A reason *for* an action is a fact in virtue of which that action is in a relevant way *better* than the relevant alternative(s).
- A reason *against* an action is a fact in virtue of which that action is in a relevant way *worse* than the relevant alternative(s).

These principles imply that, if an action is neither better nor worse in any relevant way than any relevant alternative, then there are no reasons either for or against that course of action. This seems to be a plausible implication of these principles.

What are "the relevant alternatives" here? We might be tempted to think that they are always *all* the actions that are available to the agent. But this tempting thought cannot be exactly right. Suppose that three alternatives are available to the agent—A, B, and C—where, with respect to one of the relevant values V, A is better than B, and B is better than C. Clearly, any fact in virtue of which these three acts are ranked in this way is a reason *for* A and a reason *against* C. However, if this tempting thought were correct, this fact would be a reason neither for nor against B: if the relevant alternatives to B include both A and C, then B is neither better than the relevant alternatives nor worse than the relevant alternatives—it is better than one relevant alternative and worse than another. Yet it seems that this fact must be a reason either for or against B—since it clearly bears on the question of whether or not to do B.

For this reason, we must suppose that the conversational context in which there is discussion about which actions are "good" and which are "bad", or about what is a reason "for" or "against" a given option, determines the "relevant alternatives" in a specific way. In particular, unless none of the available actions is either better or worse than any other in the relevant way, then the context must determine a sufficiently restricted set of "relevant alternatives" to each available action so that every action is in the relevant way either better or worse than all its relevant alternatives. Given our principles, this ensures that there will always be a reason either for or against every available action (at least unless none of them is either better or worse than any other in the relevant way).

As we shall see, however, it will not matter for our purposes exactly how the conversational context does this. It may be convenient in ordinary conversation to draw a line between good and bad, or between reasons for and reasons against, but if Broome's principle is correct, the most fundamental ethical theory need only focus on the *comparison* between the available actions in terms of the degree to which they exemplify each of the relevant values. It is this essentially comparative version of the value-based conception of reasons for action that is the second premise of the argument of this chapter.[2]

3. Aggregating Values

On this values-based view, then, the comparisons of the available actions in terms of the specific *reason-providing values* $\{V_1, V_2 \ldots\}$ must be capable of being somehow combined or aggregated to produce an overall ATC comparison in terms of *how much reason* there is for these actions. Indeed, fundamentally, it is not the reasons themselves that are combined or aggregated in this way; it is the specific *reason-providing values* that are combined or aggregated to determine how much reason ATC there is for all the available actions.

So far, I have just characterized values as "ways of being good". Is there a more illuminating general characterization of values that we can give? I propose that, in general, a value-concept is a concept that (a) *compares*

[2] Another account that this view of reasons can be made compatible with is that of Snedegar (2018)—although unfortunately I do not have space here to explain how to make these views compatible with each other.

or *ranks alternatives*, in a way that (b) can be expressed by terms like 'better' and 'worse', and (c) essentially plays a *reasoning-guiding role*.

By this criterion, the notion of *how much reason* ATC there is for the available alternatives is also a value-concept:

a. This concept clearly compares or ranks alternative actions, in terms of *how much reason* ATC there is for those alternatives.

b. The thought that there is ATC most reason for us to take course of action *A* can naturally be expressed by saying that *A* is ATC the "best thing for us to do".[3]

c. This notion also seems to play a role in guiding our practical reasoning, by guiding us towards *choosing* any course of action that we judge ourselves to have most reason ATC to take.

Thus, the notion of how much reason ATC there is for the available alternatives is itself a value-concept. Our question concerns how the specific reason-providing values can be aggregated to yield the overall value of how much reason ATC there is for these alternatives.

This is an importantly different conception of aggregation from other conceptions that have been discussed in the philosophical literature. The utilitarian tradition—including the "ideal utilitarianism" of G. E. Moore (1903)—tends to think of aggregation as concerning the structure of a *single* kind of goodness. More specifically, this tradition interprets aggregation as concerned with the question of how the degree of goodness exemplified by a "whole" is determined or affected by the degrees of goodness that are exemplified by the "parts" of that whole. (For example, on many versions of classical utilitarianism, the degree to which the happiness of a group of people is good is interpreted as a function—such as the *sum*—of the degrees to which the happiness of each member of that group is good.)

The conception of aggregation that I am developing here is crucially different. On my conception, for there to be any question of the relevant kind of aggregation there must be at least *three different* values that are all

---

[3] Some readers may question whether this second point is compatible with *supererogatory* actions—that is, with cases in which there is an action that is in some way *better* than all alternatives, but not such that one "ought" ATC to do it. The short answer to this question is that, in these cases, the supererogatory action is *morally* best or *most admirable*, but not in this sense "the (unique) best thing to do ATC"; on the contrary, in these cases, both the supererogatory action and some morally permissible but less admirable action are such that no alternative option is an "ATC better thing to do". For a defence and exploration of this view of supererogation, see Wedgwood (2013), and also Section 6 below.

exemplified by the members of the *same* set of alternatives. Specifically, there must be at least two specific reason-providing values $\{V_1, V_2, \ldots\}$, each of which is exemplified to some degree by each of the relevant alternatives $\{A_1, A_2, \ldots\}$. The question is how the degrees to which these alternatives exemplify these specific reason-providing values $\{V_1, V_2, \ldots\}$ determine how these alternatives compare in terms of a third *overall* value—the value of how much reason ATC there is for these alternatives—which somehow combines and aggregates the specific reason-providing values.

This idea of aggregating these specific reason-providing values into some overall or ATC value is clearly analogous to *social choice theory*. The goal of social choice theory is to explore how the preferences or utility functions of the individual members of society can be aggregated to produce a "social" preference or utility function. As I shall argue in this chapter, this is analogous to the aggregation of the specific reason-providing values into the "overall" value of how much reason ATC there is for the alternatives in question.[4]

4. Advantages of This Approach

Before developing this analogy between the aggregation of reasons and social choice theory, however, I shall canvass some considerations that support the picture of reasons that I have sketched so far. Specifically, as I shall explain in this section, this picture allows us to solve some puzzles that rival views have found extremely challenging.

On the account that was sketched above, a reason in favour of an action is a fact that *explains* why the action is better than the relevant alternatives. Now, typically, an explanation of a fact presupposes some *background* of *normal circumstances*; this background is fixed by the context in which the explanatory claim is made.

This helps to solve a puzzle due to Shyam Nair (2016: 59): "both heat and rain, taken individually, function as reasons to not run; still, the combination of heat and rain together might function as a weaker reason to not run (say, because the heat is less onerous when there is rain)." How can the conjunction of two stronger reasons result in a weaker reason in this way?

---

[4] For a pioneering discussion of the analogy between the aggregation of values and social choice theory, see Hurley (1989). For surveys of the most important results of social choice theory, see Sen (2017) and Mongin and Pivato (2016).

On my picture of reasons, this phenomenon is easily explained. Suppose that the background of normal circumstances B includes: 'Normally: if it rains, it's not hot, and if it's hot, it doesn't rain (though occasionally it rains even when it's hot).' We may think of this background as a probability distribution—specifically, a probability distribution that assigns low but non-zero conditional probability to the proposition that it's hot, given the assumption that it's raining, and also assigns low but non-zero conditional probability to the proposition that it's raining, given the assumption that it's hot.

The relevant reason-providing value seems to be indicated by Nair's allusion to how "onerous" it is to run. This value provides a *strong* reason not to run if, according to the relevant probability distribution, running can be expected to be *much* more onerous than not running. It provides a *weak* reason not to run if, according to the relevant probability distribution, running can be expected to be only *slightly* more onerous than running. In a typical context, the relevant probability function will be the result of conditionalizing this background B on the fact that is described as a "reason" in that context.

Thus, according to the probability distribution that results from conditionalizing the background B on 'It's hot', *running* is *significantly* worse than *not running* in terms of "expected onerousness". Similarly, according to the probability distribution that results from conditionalizing B on 'It's raining', *running* is again significantly worse than *not running* in terms of expected onerousness. However, according to the probability distribution that results from conditionalizing B on 'It's hot and it's raining', *running* has only *slightly* greater expected onerousness than *not running*. This explains why 'It's hot and it's raining' is a weaker reason against running than either 'It's hot' or 'It's raining'.

My picture also helps us to solve a puzzle about *counting* reasons (Fogal 2016: 88). Suppose that Ronnie loves to dance and there will be dancing at the party. Then, depending on what information is included in the background B, different statements will be true about what counts as Ronnie's reason to go to the party:

- If the background B includes 'There'll be dancing', then 'Ronnie loves to dance' is a reason (provided by the value of Ronnie's pleasure) for him to go to the party.

- If the background B includes 'Ronnie loves to dance', then 'There'll be dancing' is a reason (provided by the value of Ronnie's pleasure) for him to go to the party.
- If the background B includes neither fact, then 'Ronnie loves to dance and there'll be dancing' is a reason (provided by the value of Ronnie's pleasure) to go to the party.

However, all these three reasons are provided by the *same* value, relative to *different* backgrounds. Consequently, it would be a mistake to try to aggregate these reasons. What we fundamentally need to aggregate are the different reason-providing values—not the facts that we call the "reasons", which explain how the relevant alternatives compare in terms of these reason-providing values. Statements that presuppose *different* backgrounds create a mere illusion of different reasons. In this case, these statements are really just different ways of talking about the *same* practically significant fact—the fact about how the options of *going to the party* and *not going to the party* compare with respect to the reason-providing value of Ronnie's pleasure.

One piece of evidence in favour of my picture of reasons, as it seems to me, is the neat way in which it can solve these puzzles. Obviously, this is not a conclusive proof of this picture; but in my judgment it does give us a reason to take this picture of reasons seriously. In what follows, I shall explore the implications of this picture.

5. The Analogy with Social Choice Theory

To develop the analogy with social choice theory, a number of further premises need to be in place. These further premises fall into two groups.

First, each of the relevant values—the specific reason-providing values, and the value of how much reason ATC there is for the relevant alternatives—must be capable of being measured or represented by means of a *utility function*—and all these utility functions must be defined in terms of the *same* probability distribution.

Secondly, a pair of *Pareto* principles need to hold about the relationship between (a) the specific reason-providing values and (b) the overall value of how much reason ATC there is for the relevant alternatives. Intuitively, such Pareto principles seem highly plausible. For example, consider a case in which the only two reason-providing values are (i) beneficence and (ii) self-interest. Suppose that *A* is at least as good as *B* in terms of both beneficence *and* self-interest, and better than *B* in terms of self-interest. Then it seems plausible that there is more reason ATC for *A* than for *B*.

Once these two groups of premises are in place, an important theorem follows. Before getting to that theorem, however, I shall articulate these premises more precisely, and I shall canvass some considerations that seem to me to make these premises plausible and worth taking seriously.

i. Expected Value Measurement

First, then, we need the assumption that each of the relevant values is capable of a kind of *cardinal measurement*. Specifically, to fix ideas, I shall assume that every relevant value can be measured on an *interval* scale—that is, a scale (like the Celsius and Fahrenheit scales for temperature) on which the zero point and the unit are arbitrary, but the ratios between *differences in value* are not arbitrary. In other words, it can be literally true that the difference in value between *A* and *B* is *twice* the difference between *B* and *C*, and so on.

Next, we need the assumption that the measurement of each of these values takes an "expectational" form. According to this assumption, each of the relevant alternatives has a precise degree of this value at every *possible world* that is compatible with that alternative. However, this value is not only exemplified by these alternatives at these possible worlds; it is also exemplified by certain related *uncertain prospects*, and the value of these uncertain prospects depends on the relevant *probabilities*.

What exactly are these "uncertain prospects"? Suppose, for example, that the alternatives are *going Left* and *going Right*. Then one uncertain prospect is the act of *going Left* in all the relevant possible worlds (where one does not know for certain which of these worlds one is in); another uncertain prospect is the "mixed strategy" of *going Left* if a certain coin toss lands heads and *going Right* if it does not land heads. In general, each of these uncertain prospects corresponds to a set of possible worlds—where, for each world in the set, one performs one of the alternative actions that has a precise degree of value at that world.

If the measurement of the value takes an "expectational" form, then the value of the "mixed strategy" that I have just described is the *weighted sum* of the precise value of going Left at each world where the coin lands heads and the precise value of going Right at each world where the coin does not land heads—weighting each of these worlds by the conditional probability of the world, given the assumption that one takes this "mixed strategy". In general, the value of each uncertain prospect is the weighted sum of the precise values of the relevant alternatives at the relevant worlds, weighting each world by the relevant probability of the world, conditional on the prospect in question.

Why should it be that the measurement of these values must take this expectational form? Many different answers to this question could be suggested, but here is the answer that seems most promising to me. Perhaps it is just an essential part of each value's job description that it needs to be able to assess, not just alternatives relative to worlds, but also uncertain prospects of the kind that I have been discussing. Moreover, in assessing these uncertain prospects, the value needs to be able to get together with *any* probability function that is defined over the relevant worlds, to yield an assessment of these uncertain prospects from the standpoint of that probability function.

In our everyday use of evaluative terms like 'good' and 'bad', 'better' and 'worse', we are often willing to apply these terms to uncertain prospects, from the standpoint of a certain epistemic perspective. For example, even before knowing the jury's verdict, a defence attorney might say to her client, 'It's good that the jury has reached their verdict so quickly'. What makes this statement true is that the uncertain prospect of *the jury's reaching their verdict quickly* is rationally assessed as more valuable than the salient alternative, from the standpoint of the attorney's epistemic perspective. I suggest that we can model these sorts of epistemic perspectives by means of probability functions.

In this way, each value provides comparative assessments of the relevant prospects, from the standpoint of each probability function—assessments that could be expressed by saying such things as 'Prospect A is better than prospect B', and the like. Arguably, these comparative assessments have to meet the conditions that are imposed by the so-called "axioms" of expected utility theory.[5] (For example, these assessments must be *transitive*, and so

---

[5] More precisely, I have in mind the conditions that are imposed by the axioms of the representation theorem that Bolker (1967) proved for the decision theory of Jeffrey (1965/1983); for a lucid account of these axioms, see Joyce (1999: ch. 4). However, my approach differs from Bolker's in that the probability function P is given independently of the comparative value assessments—just as in the older proofs that were due to von Neumann and Morgenstern (1944/1953). If the probability function P is given independently, Bolker's axioms guarantee the existence of a unique utility function (given an arbitrary choice of a unit and zero point). The

on.) If these assessments must indeed meet all these conditions, then it follows that the measurement of the value must take this expectational form. In other words, the measurement of the value takes the form of a utility function.

On this approach, then, the value of uncertain prospects is relativized to the standpoint of particular epistemic perspectives—where we are assuming here that each of these epistemic perspectives can be modelled by a probability function. For example, consider the epistemic perspective of the agent who is deciding what to do. The agent can make assessments from this perspective of how the relevant alternatives compare *both* in terms of the specific reason-providing values *and* in terms of the overall value of how much reason ATC there is these alternatives; this perspective can be modelled by a probability function $P(\bullet)$. To fix ideas, let us assume that the values that we are concerned with here—at least as they apply to uncertain prospects of the kind that we have discussed—are all relativized to this epistemic perspective.

This gives us some reason to accept the following premise that we need for our argument: all the values that we are concerned with here can be measured by means of functions that have the form of *utility functions*—where all these utility functions are defined in terms of the same probability function $P(\bullet)$.

ii. The Pareto Principles

The last group of premises that I need for my argument is that the relation between the specific reason-providing values and the overall ATC value of these alternatives obeys two *Pareto principles*.

The first Pareto principle is a *Pareto Indifference* principle: if all the specific reason-providing values favour two alternatives A and B exactly *equally*, then there is exactly as much reason ATC for A as for B; A is *exactly as good* as B in terms of how much reason ATC there is for them. The second Pareto principle is a *Pareto Preference* principle: if every reason-providing

---

general idea of using the axioms of decision theory to provide a measure of goodness is due to Broome (1992: ch. 6). However, my approach differs from Broome's in two respects: (a) I do not assume that any special probability distribution is privileged as the one that is uniquely involved in the nature of the value—on my approach, comparative assessments in terms of the value can be made from the standpoint of *any* probability distribution; (b) I prefer to rely on Bolker's axioms rather than on those of Savage (1954).

value favours *A* at least as strongly as *B*, and at least one such value favours *A* *more strongly* than *B*, then there is *more* reason ATC for *A* than for *B*; *A* is *better than B* in terms of how much reason ATC there is for them.

What exactly do I mean by talking about how "strongly" a reason-providing value "favours" an alternative? As I explained in the previous subsection, all these values are relativized to the standpoint of a certain probability function *P*. So, we might think that for a reason-providing value to favour *A* and *B* equally (as assessed from the standpoint of *P*) is just for *A* and *B* to have exactly *equal* expected value, according to this probability function *P*.

If this is what we mean by how "strongly" this value favours an alternative, then the Pareto principle is what is known as an "*ex ante* Pareto principle", which encodes a strong kind of neutrality about *risk*.

For example, consider self-interest—the value of promoting one's own well-being. Suppose that one alternative *A* involves getting 100 units of well-being for certain, while a second alternative *B* involves taking a gamble between 200 and 0 units of well-being at equal odds. In this case, the expected value of *A* and *B*, in terms of how well they promote one's well-being, is exactly equal. If well-being is the only reason-providing value in this case, and the measure of how "strongly" a value favours an alternative is just the expected value of the alternative, then the Pareto Indifference principle implies that there is exactly as much reason ATC for *A* as for *B*.

It seems to me that we should avoid being committed to this strong kind of neutrality. Agents can reasonably be risk-averse, and these agents would rightly believe that they have more reason ATC for the more cautious option *A* than for the risker alternative *B*.[6] So, if we are to defend these Pareto principles, we need to adopt a different measure of how "strongly" a specific reason-providing value favours an alternative. In general, the strength with which a value favours an alternative need not be identical to the expected value of the alternative.

---

[6] I believe that a *rational* agent's choices and preferences will align with her rational expectations of how much reason ATC she has for each option. In this sense, a rational agent may be reasonably risk-averse about all values *other than* the value of how much reason there is ATC for each option. If this is right, then, for every value except for this one, we cannot derive a correct measure of the value by constructing a utility function from the agent's rational preferences between gambles involving the value: that derivation would wrongly say that the expected well-being of getting 100 units for certain is *greater* than the expected well-being of taking a 50/50 gamble between 0 and 200 units. This is why I sketched a different method for measuring the value in the previous subsection (note 5 above).

However, the strength with which a value favours an alternative relative to a world w is surely an *increasing function* of the alternative's value at w. For example, if A is *better for the agent* at w than at v, the value of goodness-for-the-agent favours A more strongly relative to w than to v; and the greater this difference in goodness-for-the-agent, the greater the difference between how strongly this value favours A relative to w and relative to v.

Still, the strength with which a value V favours an alternative A relative to a world w may reflect, not just the value of A in terms of V at w, but also some reasonable attitude towards risk. If the agent is reasonably risk-averse, then the strength with which V favours A may be a *concave* function of A's value in terms of V. As we might put it, it may be that *goodness-in-terms-of-V* makes a *declining marginal contribution* to the strength of the V-grounded reason in favour of the alternatives that are better in terms of V.

We can illustrate this point by returning to our example of the choice between getting 100 units of well-being for certain and taking a gamble between 0 and 200 units at equal odds. The difference in well-being between 0 and 100 units is the same as the difference between 100 and 200 units. But if well-being makes a declining marginal contribution to the strength of reasons, then the difference between how "strongly" the value of well-being "favours" getting 100 units and how strongly it favours getting 0 units is *greater* than the difference between how strongly it favours getting 200 units and how strongly it favours getting 100 units. On this view, we are not compelled to say that the value of well-being favours the two courses of action equally strongly—even if the expected level of well-being is the same.

With this revised understanding of how strongly a reason-providing value favours an alternative, the following Pareto principles seem plausible (in general, how strongly a value V favours an alternative A, when estimated from the standpoint of a probability function P, is just the expected strength with which V favours A according to P):

    a. If (when estimated from the standpoint of P) *every* reason-providing value V favours A *exactly as strongly* as B, then (when estimated from the standpoint of P) A is *exactly as good* as B in terms of how much reason ATC there is for them.

    b. If (when estimated from the standpoint of P) *every* reason-providing value V favours A *at least as strongly* as B, and *some* such value V' favours A *more strongly* than B, then (when estimated from the standpoint of P) A is *better than* B in terms of how much reason ATC there is for them.

Evidently, much more could be said about these Pareto principles. In what follows, however, we shall simply assume that they are correct, in order to explore their implications.

iii. The Theorem

With all these premises in place, we can now derive an analogue of John Harsanyi's (1955) "Social Aggregation Theorem".[7]

The measure of how much reason ATC there is for each relevant alternative A is a *weighted sum* of the measures of how strongly A is favoured by the relevant reason-providing values $V_1, V_2, \ldots$.

Formally, we can without loss of generality understand this theorem by focusing only on *reasons against* alternatives.

If a specific reason-providing value V does not favour *any* alternative B more strongly than A, then A is *optimal* in terms of V. In that case, there is *no* V-provided reason against A; the measure of how much V-provided reason there is *against* A is then 0. For every other alternative B, the measure $m(B)$ of how much V-provided reason there is against B is given by much *more strongly* V favours the optimal alternatives (like A) than B.

For every reason-providing value V_i, and every such measure m_i of how much V_i-provided reason there is against these alternatives, there is some reasonable way of assigning a weight a_i—a positive real number—to m_i. The measure of how much reason ATC there is against an alternative A is given by this weighted sum:

$$\sum_i a_i m_i(A)$$

The *greater* this weighted sum is, the *more reason* there is ATC *against* the alternative A. According to the interpretation of 'ought' that was canvassed in Section 0 above, an alternative is one that you *ought not* ATC to take if and only if there is ATC *more reason against it* than against some other available alternative. In this way, it is this essentially additive weighing of the reason-providing values that determines what you ought ATC to do.

---

[7] For the details of the derivation, see Broome (1992: ch. 10).

For instance, let us return to the example that we considered at the outset, of the choice between two Italian restaurants, Luna Caprese and Sole Siciliano. Suppose that there are two reason-providing values—economy and food quality.

- The value of food quality does not provide any reason against Luna Caprese at all, and it provides a *weak* reason against Sole Siciliano.
- The value of economy does not provide any reason against Sole Siciliano, but it provides a *strong* reason against Luna Caprese.

Adding the reasons against each alternative together, the combined reasons against Luna Caprese are clearly weightier than the combined reasons against Sole Siciliano. There is therefore more reason ATC against going to Luna Caprese than against going to Sole Siciliano. Going to Sole Siciliano is what you ATC ought to do.

On this conception of the aggregation of reasons, the metaphor of "weighing reasons" is not merely rough and suggestive. On the contrary, there is an *exact* parallel between the weighing of reasons and the weighing of physical quantities. On this conception, each of the reason-providing values grounds a certain quantity for each alternative—a certain amount of *reason against* that alternative; this quantity is zero if the alternative is optimal in terms of the value, and greater than zero if the alternative is suboptimal. These quantities can then all be added up, to produce a measurement of how much reason ATC there is against the alternative. Finally, we compare how much reason ATC there is against the alternative with how much reason ATC there is against every other alternative. This is exactly like weighing collections of physical objects, with the goal of selecting the lightest or least heavy of these collections.

6. Where Do the "Weights" Come From?

According to our premises, each of the reason-providing values can be measured on a unique interval scale, which is formally like a utility function. Together with some reasonable attitude towards risk, this provides a measure of how "strongly" this value "favours" the relevant alternatives. However, according to these premises, these values are all measured on *different* scales. There is no common scale on which all these values can be measured.

Thus, these weights do not come from the specific reason-providing values. Instead, they must come from a different source. It seems to me that the most plausible account is that these weights come from the standpoint of *the agent's practical reason*. In effect, they represent the way in which the agent can reasonably convert these independent measures of the specific reason-providing values into a way of measuring these values' contributions to how much reason ATC there is for the relevant alternatives. In other words, these weights express a judgment of the *relative importance* of these specific values—as measured by the measures in question—for determining what the agent ought ATC to do.

Various different views of these weights are possible here. One possible view is that there is a unique correct weighting of these measures of the specific reason-providing values, fixed by objective features of the agent's situation—such as the facts about which options are available and the values that are at stake. On this view, the agent has no discretion in weighting these measures of the values: a unique correct weighting is dictated by these objective features of the agent's situation.

On an alternative view, these objective features of the situation only fix a range of equally legitimate or permissible weightings, and there are several alternative ways in which the agent may legitimately exercise her practical reason in weighting these measures of the values in one way rather than another. On this view, then, some judgments of these values' relative importance are simply incorrect; but once these incorrect judgments are excluded, there may be a range of judgments of these values' relative importance that are all equally legitimate. It may be one mark of *saintly* agents—that is, agents who routinely perform actions that others reasonably regard as supererogatory—that they assign unusually great relative importance to moral values, compared to the value of self-interest. By contrast, less saintly agents may assign comparatively greater relative importance to self-interest. Both the more saintly and the less saintly agents may be equally reasonable—neither kind of agent is necessarily inferior to the other with respect to the virtue of practical wisdom.

In some cases, agents may be *unable* to assign any definite weights to these measures of the reason-providing values. For these agents, the conflicting values could be incommensurable. It is a good question what we should say about such cases of incommensurability, but we need not pursue this question here. At all events, such incommensurability is far from ubiquitous. In many cases, we have no difficulty in judging the relative importance of competing reason-providing values.

Undoubtedly, much further investigation is required into the questions of which ways of weighting these measures of the values are reasonable and which are not, and why these weightings are reasonable in this way. But there is no reason to assume that there is no illuminating account that could in principle be given.

7. Objections to This Additive Conception

The idea that the aggregation of reasons is essentially additive is often criticized—especially by those who are sympathetic to a "holistic" or "particularistic" view of reasons, such as that of Jonathan Dancy (2004). In fact, however, the conflict between the "Reasons Aggregation Theorem" and this holistic or particularistic view is much less stark than it may initially appear to be.

Within the framework of this theorem, the particularistic or holistic phenomena that Dancy appeals to may arise at three levels:

a. The way in which the specific values $\{V_1, V_2 \ldots\}$ of the alternatives A, $B, \ldots$ that are available to the agent in the relevant situation are determined by the *naturalistic* facts about this situation may be sensitive to the exact constellation of all these naturalistic facts.

b. The precise list of values $\{V_1, \ldots, V_m\}$ that count as "reason-providing" in the agent's situation may be determined holistically by the complete list of *all* the values $\{V_1, \ldots, V_n\}$ that are non-trivially exemplified by the available acts.

c. Finally, the precise *weighting* of the relevant measures of these values may also be determined holistically by the complete list of available acts and of the values that are non-trivially exemplified by those acts.

At the first level (a), for each of these specific values V_i, the way in which these naturalistic facts determine the degrees to which the available acts are good in terms of V_i may defy all attempts to summarize in terms of a simple principle. This will account for most of the particularistic phenomena that Dancy appeals to.

For example, consider the huge range of naturalistic facts that are relevant to determine whether an act A is an act of infidelity. It is not sufficient that if the agent does A, the agent will not fulfil a promise: it is also relevant whether the promise was given under duress, whether it was even feasible for the agent to keep the promise, and whether the promise was morally permissible in the first place. Even more naturalistic facts are relevant to determining *how much* worse, in terms of the value of fidelity, this act *A* is than the alternative acts that do not count as acts of infidelity. (Was it a solemn vow or a casual promise? How seriously is the promisee relying on the promise's being kept? And so on.)

At the second level (b), the complete list of values $\{V_1, \ldots, V_n\}$ that are non-trivially exemplified in the agent's situation may result in some of the values on this list—$\{V_n, \ldots\}$—*failing* to count as "reason-providing"—even though in other situations, in which a different complete list of values is non-trivially exemplified, these values *do* count as reason-providing. This will account for the way in which reasons can be "silenced" or "disabled" or "excluded" in the way that was famously explored by Joseph Raz (1990: 41).

For example, it may be that the value of *family loyalty*—which in many situations is a genuinely reason-providing value—may be "silenced" or "disabled" from being reason-providing in certain situations. For example, perhaps the value of family loyalty is not just outweighed, but totally silenced, when one is acting on behalf of an institution—say, on a committee that is tasked with awarding government contracts or making academic appointments. In this case, the fact that the alternatives differ in terms of the value of institutional fairness may in effect disable the value of family loyalty from counting as reason-providing at all.

The way in which it is determined which values count as "reason-providing" in the agent's situation may be important in order to avoid a kind of *double-counting*. The fact that an action would be admirable may seem to count as a reason in favour of the action. But the degree to which an action is admirable depends on the degree to which it manifests many other talents, skills, and virtues. If these virtues were included among the reason-providing values alongside the degree to which the action is admirable, the contribution that these virtues make to how much reason ATC one has for each of the available acts would in a way be counted twice over. To avoid this kind of double-counting, the list of reason-providing values needs to be determined in such a way that none of the reason-providing values depends on any of the others in this fashion.

Finally, at the third level (c), the permissible weightings of the measures of the values may depend, at least in part, on the available options and the list of values that are at stake. This clearly allows for different weightings to be permissible in different situations. Formally, it could be that in situations in which the only values at stake are V_1 and V_2, V_1 is weighted heavily, so that it normally overrides V_2, but in situations in which a third value V_3 is also at stake, V_2 comes into its own and is more often able to override V_1. In these three ways, the account proposed here can accommodate the phenomena that have motivated theorists like Dancy to opt for a holistic and particularistic approach.

I should stress that I am not endorsing the claim that any of these three holistic phenomena genuinely occur. I simply wish to emphasize that the additive conception of the aggregation of reasons is entirely compatible with all plausible claims that have been made about these phenomena.

The advantage of this conception, however, is clear. The aggregation of reasons is not an inscrutable mystery. It is a phenomenon that clearly justifies the metaphor of "weighing"—namely, an essentially additive relationship between the specific reason-providing values and the overall value of how much reason ATC there is against the available acts.[8]

References

Bolker, Ethan (1967). "A simultaneous axiomatization of utility and subjective probability", *Philosophy of Science* 34: 333–40.

Broome, John (1992). *Weighing Goods* (Oxford: Blackwell).

Broome, John (1999). *Ethics out of Economics* (Cambridge: Cambridge University Press).

Dancy, Jonathan (2004). *Ethics without Principles* (Oxford: Oxford University Press).

Finlay, Stephen (2014). *Confusion of Tongues: A Theory of Normative Language* (Oxford: Oxford University Press).

Fogal, Daniel (2016). "Reasons, Reason, and Context", in Errol Lord and Barry Maguire (eds.), *Weighing Reasons* (Oxford: Oxford University Press).

Harsanyi, John C. (1955). "Cardinal Welfare, Individualistic Ethics, and Interpersonal Comparisons of Utility", *Journal of Political Economy* 63, no. 4 (August): 309–321.

Hurley, S. L. (1989). *Natural Reasons* (Oxford: Oxford University Press).

---

[8] Earlier versions of this essay were presented at the following gatherings: a conference on Ethics and Epistemology at the University of Shandong in Jinan, China, in 2018; an Ethics Workshop at the Croatian Inter-University Centre in Dubrovnik in 2019; and the University of Arizona's Normative Ethics Workshop in 2021. I am grateful to those audiences, to Jacob Nebel, and to an anonymous reviewer for extremely helpful comments on those earlier drafts.

Jeffrey, Richard C. (1965/1983). *The Logic of Decision*, 1st/2nd edn (Chicago, Ill.: University of Chicago Press).

Joyce, James M. (1999). *Foundations of Causal Decision Theory* (Cambridge: Cambridge University Press).

Mongin, Philippe, and Pivato, Marcus (2016). "Social Evaluation under Risk and Uncertainty", in *The Oxford Handbook of Well-Being and Public Policy*, ed. Matthew D. Adler and Marc Fleurbaey (Oxford: Oxford University Press).

Moore, G. E. (1903). *Principia Ethica* (Cambridge: Cambridge University Press).

Nair, Shyam (2016). "How Do Reasons Accrue?", in Errol Lord and Barry Maguire (eds.), *Weighing Reasons* (Oxford: Oxford University Press).

Raz, Joseph (1990). *Practical Reason and Norms*, 2nd edn (Princeton: Princeton University Press).

Raz, Joseph (1999). *Engaging Reason* (Oxford: Oxford University Press).

Savage, Leonard (1954). *Foundations of Statistics* (New York: John Wiley & Sons).

Schroeder, Mark (2007). *Slaves of the Passions* (Oxford: Oxford University Press).

Sen, Amartya (2017). *Collective Choice and Social Welfare*, Expanded edn (Cambridge, Mass.: Harvard University Press).

Snedegar, Justin (2018). *Contrastive Reasons* (Oxford: Oxford University Press).

Von Neumann, John, and Morgenstern, Oskar (1944/1953). *Theory of Games and Economic Behavior*, 1st/3rd edn (Princeton, N.J.: Princeton University Press).

Wedgwood, Ralph (2009). "Intrinsic Values and Reasons for Action", *Philosophical Issues* 19: 342–63.

Wedgwood, Ralph (2013). "The Weight of Moral Reasons", *Oxford Studies in Normative Ethics*, Vol. 3, ed. Mark Timmons (Oxford University Press, 2013): 35–58.

7

The Normative Burdens of Trust

Emma Duncan

Trust is an indispensable condition of successful social functioning. It is a means of extending the efficacy of our agency and empowering the agency of others.[1] It serves to foster and deepen interpersonal relationships, allowing for the kind of secure attachments that are crucial to our development and well-being.[2] But trust is also risky. Typically, we trust when we need to count on others to help us navigate our own vulnerabilities and agential limitations. This means we must place those we trust in a position to let us down or even harm our interests. Given the riskiness of trusting and the difficulties of judging trustworthiness, it is easy to see why we are sometimes reluctant to rely on others in this special way. However, we are also sometimes reluctant for others to trust us, despite not being subject to the same vulnerabilities as trusters. If we are positioned to garner the benefits of trust without the risks that come with being a truster, what is it about the nature of trust that explains why we might reasonably resist it?

Trust theorists have recognized and attempted to address this puzzle, and their explanations have shed some light on what it is about trust that lends itself to being construed as burdensome, when it is. In trusting an agent to perform some action, we don't simply believe, desire, or hope that the person does as trusted, but we hold the person to certain *expectations*. Extant explanations of unwelcome trust highlight some important features of the central normative expectation internal to trust, but, as I will argue, if we seek a complete, robust account of trust's central normative expectation, the pictures that emerge do not yet provide one. They do, however, gesture toward a fruitful path for constructing such an account.

In what follows, I critically examine, and build on the insights of, extant explanations of unwelcome trust to motivate a novel account of the central normative expectation internal to trust. Specifically, I argue that the central

---

[1] See McGeer (2008) and Jones (2012; 2017).
[2] See Holton (1994), McCleod (2002), Wonderly (2016), Darwall (2017), and Kirton (2020).

Emma Duncan, *The Normative Burdens of Trust* In: *Oxford Studies in Normative Ethics, Volume 12.* Edited by: Mark Timmons, Oxford University Press. © Emma Duncan 2022. DOI: 10.1093/oso/9780192868886.003.0008

normative expectation internal to trust is that the trusted adopt a particular orientation of care toward the truster. This orientation not only makes sense of why trust can seem burdensome even when it is not apparently difficult to fulfill but also provides us with a view of trust's normativity that bridges explanatory gaps left by extant treatments of unwelcome trust.

This chapter is divided into five sections. In Section I, I describe some key features of the attitude of trust at issue. In Section II, I survey three common approaches to explaining what it is about the nature of trust such that it is sometimes construed as burdensome or unwelcome, where each approach appeals to a particular feature of trust's central normative expectation. In Section III, I critically evaluate the insights and limitations of these explanations. In Section IV, I argue that trust's central normative expectation consists in an invitation for the trusted to adopt an orientation of care— that is, to invest special attention in the interests and well-being of the truster in a particular domain of interaction. Finally, in Section V, I discuss some advantages of my account and address a potential objection.

I. The Attitude of Trust (and the Feeling of Betrayal)

Let us begin by specifying the type of trust at issue in this chapter. There are, after all, many ways in which we might employ the term "trust." Some uses, for example, pick out mere reliance as in "trusting" my alarm clock to wake me in time for work. Others pick out mere belief—for example, "trusting that" it will be a good day or that what was said is true. The sense of trust I mean to capture is reducible neither to mere reliance nor to belief but marks a complex attitude that we take toward other agents—one characterized by normative, though not necessarily predictive, expectations. Also note that this attitude is related to, but differs from, what we might call a "mutual bond of trust," which implies reciprocation. Although we often see such bonds in relationships like friendships or romantic partnerships, there are many cases of the attitude of trust that occur outside of these thick interpersonal relationships. Moreover, even in the context of intimate relationships, there may still be some things that we do not want to become matters of trust.[3]

---

[3] See Hawley (2014).

While there are many different accounts of trust on offer, I will, for now, work with an ecumenical conception of trust that focuses on two key marks that theorists commonly attribute to the attitude. First, trust is generally taken to be a species of reliance distinct from *mere* reliance.[4] What distinguishes trust from reliance is highly disputed, ranging from the belief that the one trusted has a commitment to do the thing trusted, to a sense of optimism regarding the goodwill of the one trusted, to normative expectations regarding the trusted's responsiveness to the fact that the truster is counting on her.[5] On some views, then, when I trust another to φ, I rely on her to φ under a certain description or from a particular perspective. If she just happens to φ for reasons wholly unrelated to her commitment, her goodwill toward me, or my dependence on her, then it's not clear that her φ-ing would fulfill my trust. An account of trust's central normative expectation that adequately explains why one might reasonably reject it should capture this key mark.

A second important mark of trust is that it makes the involved parties susceptible to certain reactive attitudes, particularly betrayal. As trusters we are susceptible to feelings such as gratitude when our expectations are met, and disappointment or betrayal when they go unfulfilled. In being trusted, we are potential targets of those attitudes and poised to experience certain self-regarding attitudes, like pride and guilt, regarding our performance. While some reactive attitudes like (moral) resentment track violations of obligations, the relationship between betrayal and obligations is less clear.

The link between trust and betrayal is widely recognized in the literature and is particularly important for those who wish to explain unwelcome trust by appealing to the type of normative expectation internal to trust.[6] But despite its import, the concept of betrayal has received little sustained attention in the trust literature.[7] And here, I can provide only a brief (but hopefully helpful) sketch that we can use going forward.

---

[4] See, for example, Baier (1986), Faulkner (2017), and Goldberg (2020).

[5] For commitment-centered accounts, see McLeod (2002) and Hawley (2014). For accounts that focus on the goodwill of the person trusted, see Baier (1986) and Jones (1996). In later work, Jones (2012; 2017) offers a dependence-responsiveness account. It is generally accepted that trust involves a belief or at least optimism about the trusted's competence to fulfill the truster's expectations, so the element that distinguishes trust from ordinary reliance tends to pertain to reasons for relying on the individual in the special way characteristic of trust. Note that not all theorists are concerned with this distinction (see Hardin (2002); Hawley (2014)).

[6] For more on the link between trust and betrayal, see Baier (1986); Holton (1994); Jones (2004); McGeer (2008); O'Neil (2012); Hawley (2014); Hinchman (2017); Kirton (2020).

[7] See Kirton (2020). For exceptions, see Shklar (1984); O'Neil (2012); and Margalit (2017).

First, the readiness to feel a sense of betrayal, as opposed to simply disappointment or resentment, is distinctively linked to the attitude of trust.[8] You may, for instance, resent an urgent care physician who negligently administers a medication to which you reported a serious allergy, yet feel both resentment and betrayal if your longtime family physician were to do the same. Both doctors ought to know better and have been remiss in their professional duties toward you. However, your inclination toward betrayal in the latter case suggests that you trust your family physician to fulfill her duties toward you but merely rely on the other.[9]

A second important feature is that betrayal does not simply track obligation or performance failures of the one trusted.[10] As in the physician case above, one may hold both parties accountable for failing to fulfill their obligations (their performance failures) but experience betrayal in response to only one. Moreover, we may feel betrayed when no obligation has been violated. For instance, I might trust a close friend to confide in me and feel betrayed when I learn they have undergone a major personal event and opted not to share it with me, while acknowledging they had no obligation to do so.

Finally, a third key feature of betrayal is that it is highly personal. The feeling of betrayal is marked by feelings of personal hurt or rejection, rather than simply resentment or disappointment.[11] This rejection often prompts us to reevaluate our relationship with the betrayer,[12] but it can also be felt deeply absent a thick interpersonal relationship.[13] Though there is scant

---

[8] See O'Neil (2012). Carolyn McLeod explains that "Feeling betrayed is the expected emotional response to broken trust, but it is not a feeling we would have toward someone on whom we merely relied" (2000: 474). See Kirton (2020) for use of the notion of betrayability in distinguishing trust from other phenomena.

[9] Theorists often describe betrayal in this context as a reactive attitude—specifically, a fitting response to violations of trust (Baier 1986; Jones 1996; Helm 2014; Darwall 2017). Notice that this attitude or feeling of betrayal is distinct from the act of betrayal. One can be betrayed without necessarily feeling the sting of betrayal. For instance, one might be betrayed by a cheating spouse yet feel relief rather than betrayal, seeing the act merely as a fortuitous excuse to dissolve an unhappy marriage. Moreover, one can feel a sense of betrayal without having been betrayed, as might be the case when one unreasonably trusts another to do something beyond their abilities and they (predictably) fail. We might helpfully characterize the act of betraying trust as disregard for the central normative expectation internal to trust, while the feeling of betrayal is a reaction to that (perceived) disregard. For an alternative view, see O'Neil (2012).

[10] Although I argue that betrayal does not track obligation, it can be appropriate in cases where obligations are flouted. Cheating on one's spouse when one is trusted to uphold their marital vows involves a violation of trust that also trades on an obligation to be faithful. For accounts that link betrayal and obligation, see Pettit (1995); McGeer (2008); O'Neil (2012); and Hinchman (2017).

[11] See Shklar (1984).

[12] See O'Neil (2012: 308) and Margalit (2017).

[13] See Kirton (2020: 592).

discussion of the exact nature of this rejection, I suggest that we can understand it in terms of a disregard for the central normative expectation internal to trust. So, to understand what is being rejected and how the truster's expectations are unmet in the relevant way, we need a more complete characterization of the central normative expectation that the trusted fails to fulfill. In the following section I explore some proposed explanations of unwelcome trust and the roles they play in illuminating the nature of trust's central normative expectation.

II. Extant Explanations of Unwelcome Trust

We can distinguish explanations of unwelcome trust in the literature roughly by the aspect of the normative expectation internal to trust that each explanation highlights. The first sort of explanation appeals to trust's normative pressure. The idea, as Karen Jones puts it, is that we may find trust unwelcome because "for one reason or another, we do not want to have to take such expectations into account, across the range of interactions the truster wants."[14] It may be that what we are trusted to do is difficult or requires a great deal of effort. Or it might be that we simply do not want to be subject to normative pressure, finding the yoke of expectation burdensome.[15] Consider the following case:

> Every day at the same time, Manny goes for a walk on the same route around his neighborhood. He does this with such regularity that his neighbors have even joked that "you can set your watch by him!" Indeed, his neighbors have come to rely on his regularity and some, on occasion, even use Manny's appearance on his route to determine the time. One day, Manny's neighbor Martha tells him that she has an important meeting the following afternoon and that she will use his walk to gauge when she should stop working in the garden and prepare for the meeting. She tells Manny that she trusts him to take his walk as usual, which Manny finds unwelcome.[16]

[14] Jones (1996: 9). Jones emphasizes the nature of trust as reason-giving, suggesting that, in a world where trust did not exist but the reasons for which others act were transparent and known to us, we still "could not have the confidence that, sometimes, another would meet [us] in [our] dependency in a domain when norm-governed and other reasons had run out for them" (2017: 100).
[15] See Darwall (2017).
[16] This case is a liberal interpretation of a characterization of Kant and his constitutionals originally described by Baier (1986) and often used in subsequent trust literature.

Although taking a walk is not difficult for Manny, he may nonetheless find Martha's trusting him to do so unwelcome because he does not wish to be subject to her expectation. This explanation implies that the attitude of trust exerts some normative pressure that one might reasonably wish to avoid. Jones has argued that trust centrally involves the expectation that the one trusted be directly and favorably moved by the dependence of the truster.[17] In other words, the fact that Martha is counting on Manny in this special way gives him a reason to do as trusted. Unfortunately, Jones says little more about how one's counting on another exerts such pressure.

Some philosophers argue that the normative pressure exerted by trust stems from the obligation it tends to confer on the trusted.[18] Others argue that trust presupposes neither the standing to make demands nor the authority to hold the trusted accountable to the truster's expectations, and, as such, its normative pressure is better understood in terms of an invitation rather than an obligation.[19] Despite disagreement about the normative structure of trust, it is widely agreed that trust opens the involved parties to certain negative reactive attitudes, suggesting another explanation for why we might find trust unwelcome.

One defining feature of the central normative expectation internal to trust is the kind of reactive attitude it leaves one open to when trust is disappointed. What I will call reactive explanations of unwelcome trust suggest that trust lays us open to certain negative reactive attitudes, and we may wish to avoid causing or being the target of these attitudes.[20] As Katherine Hawley (2014) explains, trust transforms our predictive expectations into normative ones, setting up those we trust as targets of resentment, hurt feelings, or betrayal if they fail to fulfill our expectations. Moreover, the trusted may wish to avoid being the cause of such attitudes and feelings in the truster. Thus, one might often prefer not to be trusted or, more selectively, wish to avoid certain things becoming matters of trust.

---

[17] See Jones (1996; 2012). While Jones does not explicitly characterize the expectation as normative, this seems to be a plausible interpretation of her view (see Darwall 2017: 36-8).

[18] See Hawley (2014) and O'Neil (2017). Although it is accepted that certain kinds of trust can be obligating, it is not clear that the attitude of trust itself can be. For instance, Collin O'Neil (2017) acknowledges the infelicity of taking the attitude of trust as one capable of unilaterally conferring an obligation on the trusted. We can certainly trust others to do things they are already obligated to do. And we can accept another's trust in a way that rises to the level of an agreement or promise, thereby conferring an obligation to do the thing promised. But this is not what seems to be happening in ordinary cases like Manny's where the attitude of trust is present but without what might be called a bond of trust.

[19] See Darwall (2017).

[20] See Baier (1986); Holton (1994); Jones (1996); and Darwall (2017).

The diagnosis of unwelcome trust in Manny's case, then, is that he does not wish to be positioned to betray or cause Martha to resent him if he fails to take his walk as she expects. If Martha has no authority to demand that Manny take his walk, and so he is under no obligation to do so, he may not be an apt target of Martha's resentment if he stays home, but may still yet be the cause of feelings of betrayal. So, even if he does not risk her resentment, he may still wish to avoid causing the special hurt of betrayal, and so prefer that she merely rely on rather than trust him.

Still another set of explanations of unwelcome trust centers on the content of the central normative expectation that would seem to license the reaction of betrayal. Betrayal is typically associated with close personal relationships, and some theorists posit that trust can be burdensome in virtue of the truster presuming or seeking to initiate a relationship that the trusted may want to resist. This is the central claim in what I call relational explanations. As Stephen Darwall explains, we trust "from the perspective of implied relationship with the person we trust . . ." which involves the truster either presupposing or seeking to initiate a personal relationship with the trusted (2017: 38). The relationship itself can be unwelcome, or it might require the trusted to act on relationship-specific commitments (McLeod 2002). Trust may also be unwelcome when the truster's expectation flouts the norms governing the relationship in some way (Helm 2014).

According to relational explanations, there are several reasons why Manny might find Martha's trust unwelcome. First, her trust may exceed what she can reasonably expect of Manny given the norms governing their relationship. Second, if her trust is only appropriate given the norms of a relationship that doesn't obtain between Manny and Martha, or Manny does not desire the sort of relationship presupposed, he may find her trust unwelcome. Finally, fulfilling Martha's trust may require Manny to act on commitments that are not characteristic of the sort of relationship Manny has (or wants) with Martha.

Each kind of explanation reveals something important about trust's central normative expectation. Explanations that appeal to the normative pressure exerted by trust show that, whatever its normative structure, the attitude of trust has some normative force for the trusted. Reactive explanations show that the truster's expectations leave the involved parties open to certain reactive attitudes, particularly betrayal, which is a unique response to the disappointment of those expectations. Finally, relational explanations suggest how interpersonal relationships between the truster and the trusted can affect the appropriateness of the truster's expectations. However, while these explanations offer important insights about the central normative expectation internal to trust, they also highlight the need to address certain explanatory gaps and unanswered questions.

III. Limits of Extant Explanations

Above, I argued that extant explanations of unwelcome trust could help inform the nature of trust's central normative expectation. Employing insights gleaned from these explanations, I will now consider a case that highlights some important remaining questions about trust's central normative expectation.

Bothered Bartender

A man walks into a bar. Having just come from retrieving his father's cremated remains, he places an urn on the bar top beside him, orders a drink, and strikes up a friendly conversation with the bartender, Joe. Eventually, Joe goes to the back for supplies. Upon his return he finds the man gone, having left his urn and a note. The note explains that the man had to run an important errand that could take a few hours, and it was not practical to take the urn with him. He says he trusts Joe to look after it and thanks him for doing so. Although Joe's shift will last several more hours and he can easily store the urn behind the bar, he nonetheless finds the man's trust in him unwelcome.

One explanation for Joe's discomfort can be found in the normative pressure exerted by the kind of expectation internal to trust. Proponents of this explanation might argue that Joe finds the man's trust unwelcome because he does not want to be subject to his expectations. The problem is that it is unclear from this explanation what it is about trust that exerts the relevant pressure.

One might reply that the central normative expectation internal to trust has the form of a demand and confers an obligation on the trusted, and this is what Joe finds unwelcome. However, we often trust others to do things we have no standing to demand of them, and it is not clear what it is about the man's holding an attitude of trust toward Joe that could be obligating. Even if we represent Joe as having a professional obligation that the man trusts him to fulfill, it is not the man's trust that exerts the relevant pressure but Joe's pre-existing obligation, and it is not clear why Joe would suddenly find that obligation unwelcome because the man now trusts him to fulfill it.[21]

Finally, one might argue that if Joe were to accept the man's trust, his acceptance would constitute an agreement or promise that confers an obligation, and this explains what Joe finds unwelcome.[22] This seems plausible, though it would not be the man's trust that binds but the agreement or promise. The mere attitude of trust will not suffice to ground an obligation. What Joe would find unwelcome, then, is not the man's trust *per se* but the obligation stemming from Joe's acceptance of his trust. But perhaps there is another aspect of trust's central normative expectation that explains why Joe finds the man's trust unwelcome, namely that it makes Joe a potential target of the man's negative reactive attitudes.

Whatever the normative structure of trust, Joe might be uncomfortable with the prospect of being the cause or target of the man's resentment or feelings of betrayal if Joe should fail to fulfill his trust. Although such reactive explanations highlight a potentially burdensome feature of trust, they have some difficulty accounting for the subset of reactive attitudes typically associated with trust. First, although it may sometimes be appropriate to resent the trusted for failing to fulfill our trust, moral resentment seeks accountability for flouting moral obligation and trust does not obviously obligate. As Stephen Darwall notes, failures of trust "do not, in themselves, justify resentment and blame, so much as other more personal responses like being 'let down' or some other form of personal hurt" (2017: 40). But it is not clear what it is about trust that warrants this sort of feeling. For instance, if the central normative expectation internal to trust is that the trusted is directly and favorably responsive to the dependence of the truster (Jones 1996), it is not clear why the appropriate reaction is hurt feelings rather than, say, frustration. So, explanations that appeal to our aversion to being the target (or cause) of negative reactive attitudes must still account for trust's link to a certain subset of attitudes, including betrayal.

Further, given the unique link between trust and betrayal, reactive explanations must speak to what it is about the nature of trust that explains why betrayal is sometimes an appropriate reaction to disappointed trust. As

---

[21] See O'Neil (2012) for more on trusting others to fulfill pre-existing obligations.

[22] One might also argue that in the context of intimate relationships, which are pervaded by a bond of trust, such agreements (and therefore obligations) abound. I will address this in more depth later, but even in such cases I suspect that the obligations are grounded by relationship norms rather than the attitude of trust.

Collin O'Neil notes, even when one takes the trusted to be under an obligation, it is the readiness to feel betrayal rather than resentment or mere disappointment, in response to failures by the trusted, that is distinctive of the attitude of trust (2012: 308). It might be appropriate, for instance, to resent a stranger (whom you do not trust) for stealing your laptop but not to feel betrayed, whereas it does seem appropriate to experience betrayal (as well as resentment) if the theft is committed by a friend. One explanation is that, in cases of betrayal, the trusted manifests a failure to engage the needs of the truster in the way expected (Hinchman 2017: 51). But this sort of explanation calls for a more specific articulation of trust's central normative expectation in order to account for the differences in expectations between the friend and the stranger. Favorable responsiveness to the dependency of the truster and acting in accordance with one's obligation seem too general to play this role. Relational explanations, however, offer a more promising path to the sort of articulation we seek.

Relational explanations appeal to relationship-specific norms to explain what is personal, and sometimes unwelcome, about trust. For instance, Darwall explains that what trust seeks is the kind of mutual recognition that is distinctive of personal relationships like friendship and love, which we have no authority to demand, but hope for and feel hurt or let down when unreciprocated (2017: 46). This explains trust's connection to personal reactive attitudes rather than accountability-seeking attitudes like resentment. It also helps explain what we might find unwelcome, namely the second-personal recognition and responsiveness characteristic of a kind of close relationship that might not obtain between the truster and trusted. This may be particularly problematic since part of what trust seeks is a kind of reciprocity (Darwall 2017). For Darwall, although trust does not issue a demand, it does come with an implicit RSVP, a call for the trusted to hold some fitting attitude in return (2017: 40–1). As Darwall explains,

> ...it does seem essential to my trusting you that I invite you to accept my trust and, indeed, that I invite you to trust that I am indeed trusting you, to trust in my trust and in me, trusting you. It will turn out that trust is a reciprocating attitude to itself. (2017: 42)

Importantly for explaining unwelcome trust, the second-personal standing of the truster in this call-and-response structure can be felt as an 'imposition' by the trusted (Darwall 2017: 40). So, in Joe's case, the man's trust may be unwelcome, not because it is obligating, but because it calls on him to respond to the offer of trust in a way that presupposes or seeks to initiate a relationship Joe may not desire. A central expectation of the man's trust is that Joe reciprocate with trust, something Joe may feel is unwarranted or would make him (qua truster) vulnerable to the man in undesirable ways.

But it is not at all clear that this is what we expect when we trust. After all, it makes sense to trust strangers without expecting that they trust us in return. Moreover, although trust can be integral to building or reshaping relationships, it does not obviously presuppose any particular sort of relationship that we may reasonably wish to resist.

One might reply here that the difficulty lies in the truster's expectation that the trusted act on a relationship-specific commitment (McLeod 2002: 31–3; Hawley 2014). Different kinds of relationships are partly distinguished by the sort of concern and commitments that comprise them. A call to have what Carolyn McLeod labels 'special concern' for another, beyond the specific concern we are committed to having toward everyone, can be unwelcome. If a stranger's trust calls on you to act on a commitment you only have in a close friendship, that call may be unwelcome, especially if it requires you to reinterpret the nature of your relationship to make sense of acting on the relevant commitment.[23] As McLeod explains, "...trust is unwelcome when the trusted one does not perceive her relationship with me as the kind that requires her to fulfill the responsibility I trust her to fulfill" (2000: 171, note 4). Similarly, Bennett Helm argues that trust is unwelcome when it fails to root the responsiveness sought by the truster in respect for the community's (relationship-governing) norms (2014: 208). That is, insofar as the truster's expectations exceed what it is reasonable to demand given her relationship with the trusted, the trusted may find them unwelcome.

Although the norms and expectations governing the relationship between the truster and the trusted can go some distance in explaining why we might sometimes find trust unwelcome, some unanswered questions remain. First, as noted earlier, the attitude of trust does not issue a demand, and we often trust others to do what we cannot demand of them, so it is not obvious why the attitude of trust violates any norms in Helm's sense. Second, although we sometimes trust others to fulfill their commitments, in many cases what we

---

[23] While acting on one's commitments is central to McLeod's view, according to which the motivation expected from the trusted is moral integrity, Hawley (2014) holds that one may act only in accord with, though not necessarily on, one's commitments and still fulfill the truster's expectations.

trust others to do is not tied to any specific commitment, let alone one that is distinctive of a particular relationship. I may have a commitment only to close friends that I help them move, but I can help an acquaintance move without signifying that I hold such a commitment to them or that they should reinterpret the nature of our relationship to include one.

Finally, one might argue that if the central normative expectation internal to trust is that one will act with moral integrity to fulfill their commitments, the violation of these commitments, especially in the context of thick interpersonal relationships, accounts for the personal character of betrayal as a response to broken trust (McLeod 2000). However, without an explanation of the personal (rather than merely moral) significance of fulfilling such commitments it is not clear why their violation would elicit personal feelings like betrayal rather than a moral evaluation of the trusted. This explanatory gap suggests that, even within the rich normative context of thick interpersonal relationships, we need more to fully account for trust's personal character.

Thus, if we wish to glean a more complete understanding of trust's central normative expectation from an explanation of unwelcome trust, the explanation must make sense of the normative pressure exerted by trust, the distinctively personal reactive attitudes associated with trust, and how relationship considerations can affect trust.

IV. Trust and Care

In this section I sketch dual proposals for (i) an explanation of unwelcome trust and (ii) an account of the central normative expectation internal to trust. Specifically, I suggest that trust invites the trusted to adopt a particular orientation of care toward the truster, which the trusted might reasonably find unwelcome. I elaborate on the relevant features of invitation and care in what follows.

While invitations lack the peremptory force of moral demands, they nevertheless exert pressure for discursive uptake on the invitee.[24] Specifically, they call on the invitee to recognize and respond to the invitation by accepting it or

---

[24] See Telech (2020). By "invitation" I mean to characterize the type of "call" internal to trust. I mean to suggest neither that the invitation must be explicitly expressed nor that the truster must intend her trust as an invitation. For instructive discussion of how attitudes can function as forms of moral communication in the relevant sense, see Macnamara (2015).

providing some reason for not doing so (Darwall 2017). For example, if you invite me to your party, you have not demanded that I attend and I do not feel as though I *must* go, but I will feel some pressure to attend or provide an excuse. Unlike Darwall, who construes trust as both an invitation to accept trust and an invitation to the trusted to reciprocate with trust, I argue, in line with Jones (2012; 2017), that what trust seeks largely consists in responsiveness to the dependency of the truster. However, parting from Jones, I argue that what the truster seeks is not mere responsiveness to dependency but care—specifically, for the trusted to make a particular domain of the truster's interests a special locus of her attention.[25]

Theorists tend to characterize caring in terms of a suite of psychological states and dispositions which includes patterns of emotion, judgment, and attention. On Helm's view, for example, genuine care requires "a consistent pattern of attending to the relevant object: in short, a kind of *vigilance* for what happens or might well happen to it" (2010: 57). Agnieszka Jaworska describes caring as "a structured compound of various less complex emotions, emotional predispositions, and also desires, unfolding reliably over time in response to relevant circumstances" that construes its object as a source of import (Jaworska 2007: 560). Building on Jaworska's insights, Jeffrey Seidman argues that caring consists of "a disposition to attend to an object and hence to considerations pertaining to it, and a disposition to respond to the real or apparent reasons those considerations provide" (2016: 2785). These descriptions will be helpful in elaborating the caring orientation I take to be internal to trust.

Like (some versions of) the thicker notion of care described above, the relevant orientation of care is non-instrumental (directed at the object for its own sake), shapes one's deliberative boundaries, and is constituted by dispositions to attend to the object of care, to be emotionally vulnerable to how it fares, and to see it as a source of reasons for action and emotion.

Although the caring orientation invited by trust is less demanding than some of the richer and more robust conceptions employed by some caring theorists, it is typically more demanding than the bare concept of goodwill

[25] That trust calls for some sort of "care" is a familiar notion in the literature. Baier (1986) introduced the idea that trust is reliance on the trusted's goodwill toward the truster, and Jones (1996) echoed this, though she later argued that the concept is vague and often stretched beyond useful meaning (2012). Perhaps most similar to the kind of care I have in mind is McLeod's "special concern" (2002). As will become clearer in what follows, the orientation I describe is more extensive than goodwill and not tethered to relationship-specific commitments, distinguishing it from both Baier's and McLeod's conceptions.

sometimes found in theories of trust.[26] The relevant caring orientation is more of a psychological investment, constituted by the suite of attitudes and dispositions described earlier, like emotional vulnerability to the truster's interests, which goodwill needn't involve. For instance, although being positively responsive to another's dependency is a way of showing goodwill (Jones 2012: 69), one may show goodwill in other ways without the normative pressure to act. Goodwill can manifest in my hope that a stranger's unattended bag will not be tampered with, though I do not feel pressure to pay it special attention as I would were I to take a caring orientation toward the stranger's interests.

Furthermore, the caring orientation called for by trust can license inferences and further expectations that the, often vague, notion of goodwill does not. Goodwill plays a different relationship-oriented role than care. On many accounts, we are regularly required to display goodwill toward our fellow moral agents, whereas trust seeks a response that not all others are normally expected to take toward us. Given its closeness to the kind of caring attitudes that form the glue of more intimate relationships, even this more penumbral caring orientation can create pathways for initiating and advancing relationships in ways goodwill cannot. This makes sense of why we afford trust an elevated status in our interactions with others. We often accept it, in all its risks, recognizing that part of what makes it a special orientation is that it does carry some normative burdens of the sort we might sometimes wish to do without.

To see these features of the caring orientation at work, consider a situation in which a stranger at a café trusts you to look after her laptop while she steps away to take a call. In trusting you, the stranger invites you to devote special attention to her interests, for her own sake. This entails the safekeeping of her laptop because doing so matters to *her*. The orientation of care called for by trust involves making the truster's interests a special locus of attention for the trusted, rather than it merely turning out that the truster's interests overlap or coincide with those of the trusted.[27] For example, if you watch the laptop only because you intend to steal it for yourself later, then you are not fulfilling the stranger's trust. Similarly, if the stranger thinks you will watch the laptop solely because you fear that a pattern of thefts will sully the reputation of your favorite café, she may rely on you, but not trust you, to do it. Furthermore, attending to her interests in

---

[26] See Baier (1986) and Jones (1996).
[27] See Hardin (1993; 2002) for an alternative view.

the relevant way shapes your deliberative boundaries, ruling out some options and making others more salient. Whereas you might normally be inclined to leave the café, having finished your work, departing before the stranger returns won't be a live option. Or, you might notice, and attempt to correct, the laptop's unstable positioning on the table so it doesn't fall. The orientation also involves motivations to act in ways (reasonably) required to fulfill the truster's expectation, and dispositions toward certain self-directed reactive attitudes regarding your performance. Bound up in this orientation is also a disposition to be emotionally vulnerable to how the truster's interests (in the security of the laptop), which you have made an object of import for you, fare. If, for example, your attention lapsed and the laptop was stolen, you should be inclined to feel a degree of sadness or regret at the significance of the loss *for her*, not as a loss to you, nor as mere disappointment or guilt at your failure to live up to her expectations.[28]

How does this caring orientation help us understand why trust is sometimes unwelcome? First, the invitational structure of trust generates normative pressure for discursive uptake by the addressee. The addresser asks for a fundamentally important and personal orientation, the denial of which often results in hurt feelings. This can make even polite rejections of the truster's expectation uncomfortable and position us to hurt another's feelings. Failing, or refusing even to try, to adopt an invited caring orientation may also seem like a personal slight toward the truster.

Second, understanding the central normative expectation as asking another to adopt a particular orientation of care one has no standing to demand also helps explain why the subset of reactive attitudes associated with trust is so personal. We recognize that often when we trust we expect the trusted to respond in a way that we are not owed but has deep import and personal meaning. The orientation we seek as trusters is rooted in many of the same inclinations and dispositions that characterize intimate relationships and has special significance when it is freely given, especially in the absence of a relationship that calls for it. While this significance helps explain the personal nature of the specific reactive attitudes associated with trust, particularly betrayal, it also helps explain how relationship norms affect the appropriateness of trust, even between non-intimates.

The orientation of care invited by trust is of a similar, albeit penumbral, form of the orientation characteristic of more intimate relationships. The

---

[28] This isn't to deny that self-regarding disappointment, or even guilt or shame, are also aspects of the relevant emotional vulnerability.

trusted, then, might reasonably worry that adopting this orientation could falsely convey to the truster that the pair have (or could come to have) a closer relationship than the trusted desires. In the bartender case, for example, Joe may have been happy to look after the urn as part of his role as a bartender but is reluctant to represent himself as having the sort of orientation toward the man that could be construed as grounding a relationship other than that of bartender and patron. But we may find trust unwelcome for an even more personal reason.

Given the thicker nature of the orientation I have described, we may not want to adopt it toward just anyone at any time. It is not simply a matter of wishing to avoid being subject to personal reactive attitudes; it is that we may wish only to be so oriented in certain cases. For instance, Joe may be reluctant to attach special import to the interests of a stranger with whom he has no special connection, though he would be willing to accept the man's reliance on him along with attributions of the kind of goodwill we can reasonably expect from others. This is not simply in virtue of the role the orientation plays in undergirding relationships but is partly due to what is involved psychologically in taking up such an orientation. Considering the personal nature of its central normative expectation, then, it is no surprise that we are sometimes reluctant to accept the trust of others—to make what matters to them matter to us.

V. Advantages and Objections

Before examining a potential objection to my view, it is worth reviewing some advantages of my proposed account of trust's central normative expectation. First, it explains the normative pressure exerted by the attitude of trust in terms of the content and structure of the central normative expectation internal to trust. Second, the orientation of care explains the link between trust and the personal reaction of betrayal that reactive accounts identify as a distinctive burden of trust. Recall that the hurt felt in betrayal seems to stem from the sense that one's interests do not matter sufficiently to the trusted. This is the heart of the sense of abandonment and rejection sometimes felt when another conveys a lack of care about what matters to you, even when they face no obligation or commitment to care. When such care is partly constitutive of a relationship there might be a normative expectation to hold such a commitment or even an obligation to do so. This helps explain why, while betrayal can be an apt response to failures by non-intimates or strangers, it is a much more common reaction to failures of trust within thick relationships. Still, even in these relationships there may be certain things we wish not to become matters of trust, precisely because we don't want *those things* to signify the nature and degree of care we have toward another.[29]

The caring orientation also explains, whereas reactive accounts do not, why trust in particular might be unwelcome when nearby phenomena like hope or mere reliance, which also risk hurt feelings, are not. Identifying betrayal as a reactive attitude unique to trust is not sufficient to account for this difference without an accompanying characterization of the normative expectation internal to trust that links it to betrayal.

Finally, although the relational views I have examined hit close to the mark, their construals of the relevant expectation rely on elements that are either too thin, impersonal, or narrow to fit the phenomenology of trust. The caring orientation I have described captures the personal nature of trust and its characteristic reactive attitudes better than responsiveness-based and respect-based explanations and without imposing an overly demanding view of the expectations internal to trust—for example, acting on relationship-specific commitments or trusting the truster in return.

One might worry that the orientation I have described asks too much of those we trust, particularly when dealing with strangers. It either makes trust so demanding that it appears undesirable without an antecedent relationship whose health and maintenance call for it, or it makes trust between strangers too rare. One might think it makes little sense to accept the trust of a stranger who asks you for directions if their trusting you requires you to make their interests a special locus of attention and to make yourself emotionally vulnerable to how those interests fare. Moreover, the apparent demandingness of this orientation does not fit with our everyday experiences of trusting strangers. When we trust a stranger for directions, we do not take ourselves to hold the expectation that they adopt a caring orientation toward us—it seems that what really matters is their competence and the likelihood that they are telling the truth, or so the objection goes. There are two avenues of reply here.

First, in ordinary language, we are not always careful to distinguish trust from mere reliance. Given our promiscuity with the term, we might mistakenly use "trust" where "reliance" would do better. Many instances of one-time interactions with strangers may more aptly be described as reliance

---

[29] Hawley makes a similar point using an example of one romantic partner trusting another to do a specific household chore (2014).

rather than trust, such as when we casually speak of trusting a concierge to recommend a great seafood restaurant nearby. If the recommendation is a bust, we do not tend to feel betrayed, but instead feel frustrated or irritated that they didn't do their job well. Similarly, when we speak of trusting drivers to obey the rules of the road, the situation seems better construed as one in which we rely on others to respect traffic laws and the moral demand not to recklessly endanger others. Again, we are more apt to feel resentment than betrayal when a driver cuts us off.

Sometimes, though, we do trust strangers in the relevant sense—inviting them to adopt a caring orientation toward us. Consider a case: you leave your optometrist's office after having your pupils dilated, and with your vision still blurry you need to cross a busy intersection. One of the strangers at the crosswalk notices your dark glasses and sympathizes with your situation, having been in a similar one herself. Her sympathetic interest in your situation might be enough to signal to you that you could rely on her in a special way to get you across the street. It might be reasonable to expect, for example, that she take extra care to point out obstacles in the crosswalk or not rush across and leave you behind, despite not actually having agreed to assist you. Although signaling trustworthiness is a complicated matter,[30] the point is that if you do take her engagement with you as a signal of trust-worthiness, it is because it signifies that she has or is willing to adopt the kind of orientation sought by trust. For instance, if after your banter you began to follow her and she rushed ahead, leaving you to fend for yourself, you might feel betrayed, though you wouldn't if the other strangers had done the same—you have *trusted* her.

The second reply is that while the caring orientation involved in the optometry case certainly was special, not one that can simply be expected from just anyone, it is not obviously too demanding to be appropriate given the relationship between the truster and trusted. Familiarly, caring comes in degrees and forms, some of which are richer and deeper than others. While caring of this sort can lay the groundwork for developing relationships, it needn't constitute or initiate a lasting relationship.[31] We are capable of adopting (and commonly do) an orientation of care toward others with whom we have little or no antecedent relationship, making their interests in

---

[30] See Jones (2012; 2017).

[31] For example, we may exhibit a caring orientation when engaged in joint activities, like taking a walk together, and feel normative pressure to help the other see the activity through, without the orientation or sense of commitment persisting beyond that venture (see Gilbert 1990: 6).

a particular domain a special locus of attention for us for a time, without committing ourselves to that orientation or a more substantive relationship.

Conclusion

One benefit of the account I have offered is its ability to tie together the apparently disparate insights from extant explanations of unwelcome trust to form a clearer picture of the content and shape of the central normative expectation internal to trust. If the normative burdens of trust indeed stem from the caring orientation I claim that trust seeks, it helps us understand (i) what it is about trust, as opposed to hope or mere reliance, that exerts normative pressure that is sometimes unwelcome, (ii) how it is that distinctly personal reactive attitudes like betrayal are appropriate reactions to violations of trust, and (iii) that a central feature of relationships, namely the orientation of care, is embedded in trust's expectations such that trust can advance and even initiate relationships in ways that might be unwelcome under certain conditions.

I have argued that an examination of why trust is sometimes construed as unwelcome reveals that the core of trust's expectation is an invitation for another to take a caring orientation toward the truster, to make her interests in a particular domain a special locus of attention. In trusting, we ask others to make what matters to us matter to them, for our sake. This orientation makes sense of claims that trust can sometimes be felt as burdensome, in a way that explains the uniquely personal nature of the reactive attitudes characteristic of trust, and illuminates the dynamic interplay between trust and interpersonal relationships.[32]

Works Cited

Baier, Annette. "Trust and antitrust." *Ethics* 96, no. 2 (1986): 231–60.

Darwall, Stephen. "Trust as a Second-Personal Attitude (of the Heart)." In *The Philosophy of Trust*, 35–50. Oxford: Oxford University Press, 2017.

---

[32] Many thanks to David Brink, Rosalind Chaplin, Coleen Macnamara, Dana Nelkin, the participants at the 2021 Arizona Workshop in Normative Ethics, and two anonymous reviewers of *Oxford Studies in Normative Ethics* for their very helpful comments on this chapter. Special thanks to Monique Wonderly for her instructive feedback and many insightful conversations about the nature of trust.

Faulkner, Paul. "The problem of trust." In *The Philosophy of Trust*, 109–28. Oxford: Oxford University Press, 2017.

Gilbert, Margaret. "Walking together: A paradigmatic social phenomenon." *Midwest Studies in Philosophy* 15 (1990): 1–14.

Goldberg, Sanford C. "Trust and Reliance." In *The Routledge Handbook of Trust and Philosophy*, 97–108. New York: Routledge, 2020.

Hardin, Russell. "The Street-Level Epistemology of Trust." *Politics & Society* 21, no. 4 (1993): 505–29.

Hardin, Russell. *Trust and Trustworthiness*. New York: Russell Sage Foundation, 2002.

Hawley, Katherine. "Trust, Distrust and Commitment." *Noûs* 48, no. 1 (2014): 1–20.

Helm, Bennett W. *Love, Friendship, and the Self: Intimacy, Identification, and the Social Nature of Persons*. New York: Oxford University Press, 2010.

Helm, Bennett W. "Trust as a Reactive Attitude." *Oxford Studies in Agency and Responsibility* 2 (2014): 187–215.

Hinchman, Edward S. "On the Risks of Resting Assured: An Assurance Theory of Trust." In *The Philosophy of Trust*, 51–69. Oxford: Oxford University Press, 2017.

Holton, Richard. "Deciding to Trust, Coming to Believe." *Australasian Journal of Philosophy* 72, no. 1 (1994): 63–76.

Jaworska, Agnieszka. "Caring and Internality." *Philosophy and Phenomenological Research* 74, no. 3 (2007): 529–68.

Jones, Karen. "Trust as an Affective Attitude." *Ethics* 107, no. 1 (1996): 4–25.

Jones, Karen. "Trustworthiness." *Ethics* 123, no. 1 (2012): 61–85.

Jones, Karen. "But I was counting on you!" In *The Philosophy of Trust*, 90–108. Oxford: Oxford University Press, 2017.

Kirton, Andrew. "Matters of Trust as Matters of Attachment Security." *International Journal of Philosophical Studies* 28, no. 5 (2020): 583–602.

Macnamara, Coleen. "Reactive Attitudes as Communicative Entities." *Philosophy and Phenomenological Research* 90, no. 3 (2015): 546–69.

Margalit, Avishai. *On Betrayal*. Cambridge: Harvard University Press, 2017.

McGeer, Victoria. "Trust, Hope and Empowerment." *Australasian Journal of Philosophy* 86, no. 2 (2008): 237–54.

McLeod, Carolyn. "Our Attitude Towards the Motivation of Those We Trust." *Southern Journal of Philosophy* 38, no. 3 (2000): 465–79.

McLeod, Carolyn. *Self-trust and Reproductive Autonomy*. Cambridge: MIT Press, 2002.

O'Neil, Collin. "Lying, Trust, and Gratitude." *Philosophy & Public Affairs* 40, no. 4 (2012): 301–33.

O'Neil, Collin. "Betraying Trust." In *The Philosophy of Trust*, 70–89. Oxford: Oxford University Press, 2017.

Seidman, Jeffrey. "The Unity of Caring and the Rationality of Emotion." *Philosophical Studies* 173, no. 10 (2016): 2785–801.

Shklar, Judith N. *Ordinary Vices*. Cambridge: Harvard University Press, 1984.

Telech, Daniel. "Praise as Moral Address." *Oxford Studies in Agency and Responsibility* 7 (2020): 154–81.

Wonderly, Monique. "On Being Attached." *Philosophical Studies* 173, no. 1 (2016): 223–42.

8

Attributive Silencing

Mark Schroeder

My aim, in this chapter, is to explore a novel form of silencing first discussed in print by Mary Kate McGowan (2017), which I call *attributive silencing*. Whereas Rae Langton (1993), Jennifer Hornsby (1995), and others have brought the tools of analytic philosophy of language to bear in helping us to understand ways in which people can be rendered unable to be successful in communicating, and Kristie Dotson (2011) and others have explored how epistemology can reveal other ways in which this can happen, we learn about the possibility of attributive silencing through the philosophy of action and the theory of responsibility. Attributive silencing, I will argue, far from a bare theoretical possibility, is an indisputably genuine phenomenon, but how easily it can happen will depend on the correct theory of attributability, and hence on the correct theory of persons. On some theories of persons it will be rare, but on my preferred, *interpretive* account of persons attributive silencing can happen quite easily and without fault. The phenomenon of attributive silencing therefore has many important possible applications, but I will close by briefly considering how it applies specifically to women's refusals of sexual advances, one of the original applications of theoretical work about silencing.

1. Background

wIt is the holiday season as I write, and shopping centers everywhere are filled with the sound of seasonal music. Classics sung by Bing Crosby and new additions from Taylor Swift mix with the obligatory Mariah Carey and Kenny G. But you can never go far without encountering one of the over two hundred (according to Wikipedia) different recordings of 'Baby It's Cold Outside'.[1]

---

[1] I can't vouch for the accuracy of this number; Wikipedia only catalogues seventy-one distinct recordings.

Mark Schroeder, *Attributive Silencing* In: *Oxford Studies in Normative Ethics, Volume 12*. Edited by: Mark Timmons, Oxford University Press. © Mark Schroeder 2022. DOI: 10.1093/oso/9780192868886.003.0009

Perennial fodder for social media disputes over whether it is a cute depiction of what counted for sexy flirting in the 1940s or an exemplification of rape culture that glamorizes lack of sexual consent, the lyrics consist in an exchange between a female voice who says things like 'I really can't stay…I gotta go away', 'my father will be pacing the floor', and 'say, what's in this drink?', while the male voice responds, 'baby, it's cold outside' and 'how can you do this thing to me', and encourages her to stay for one more drink.

The striking feature of this song on which I want to focus is that such different interpretations of the same conversation are even possible. On one interpretation—call it the 'intended' interpretation—a woman wants to stay and have a drink but is worried about appearances, and so must put up the appearance of resistance in order to exercise her own sexual agency. Whereas on another interpretation—call it the 'available' interpretation— the woman really is trying to get away while making polite excuses, but her excuses are being rejected as if they are her only reasons, rather than the making-polite way of expressing rejection. No doubt the intended interpret- ation is aptly named, but what interests me and should interest all of us is the fact that the very same text could be a script for either dialogue, and interpretive judgment is required in order to understand which script is being followed.

All of us, I think, have at some point or another had the experience of trying to tell someone something but being unable to get through to them. This is the phenomenology of being silenced—or at least, of *some ways* of being silenced. If you have not yourself been in the difficult position of trying to let someone's romantic or sexual interest down lightly, you may at least have teenage children to whom it is difficult to get through, or—more likely—have been a teenager with parents to whom it is difficult to get through. But decades of work in feminist theory have made clear that the experience of being silenced is not equally distributed. It is a much more dominant and prevalent part of the experiences of those belonging to groups subject to various forms of oppression and marginalization. The controversy over 'Baby It's Cold Outside' helps all of us—regardless of the force of our own encounters with this experience—to hold firmly in view the reality of this phenomenon. The experience of silencing is real, and silencing is recognizable—the only questions are *how* it is possible, and what is actually happening in cases like this.

This is where philosophy comes in. And here it will help to regiment vocabulary slightly. Let us say that someone is *silenced* in the respect, and to the degree, that they face some obstacle to communicating what they would like to communicate. In so regimenting this term for the purposes of this chapter, I am following Maitra (2009) in setting aside some of Langton's (1993) illustrative examples of illocutionary disablement in which the illocutionary act that someone is prevented from performing serves some non-communicative purpose, such as marrying or naming. But I leave open the possibility that silencing can come in degrees. And I purposely leave open whether the obstacle that they face has been imposed by anyone who *does* the silencing. We can still say that some person *silences* another, treating 'silence' as a transitive verb, but I just insist that we leave open whether someone can be silenced without anyone silencing them, and I leave open whether only persons belonging to oppressed classes can be silenced—with of course the expectation that this will turn out to be the paradigm.[2]

I will also distinguish the state of being silenced from the experience *as of* being silenced. Often, of course, when someone has an experience as of being silenced, they really are silenced. I don't doubt that this is true, and everything that I say in what follows will go towards confirming it. But, strictly speaking, this is not logically established by any of our real experiences as of being silenced. And we should leave open the possibility that not all experiences as of being silenced are veridical. For example, if the teenager feels unable to get through to their parents and their parents feel unable to get through to the teenager, then it may help to reserve judgment on the possibility that one of their experiences is non-veridical—certainly each can acknowledge that the other has an experience as of being silenced without giving up the game and granting that this experience is veridical.

So understood, silencing could come in many forms, due to many different possible obstacles to someone communicating what they would like to communicate. The simplest such form, and the easiest to understand, is that someone can face disincentives for speaking their mind, so that they do not speak out at all. Generalizing slightly, let us call such

---

[2] Some will prefer to reserve the term 'silencing' for use in contexts of oppression, but that can't settle the substantive question of whether silencing as I have understood it can also happen in the absence of independent forms of oppression. So from an analytic perspective it is helpful to use a term that leaves this question open (compare Caponetto (2021)). As I have already insisted and will remain clear, I agree that even if we do not stipulatively narrow our use of 'silencing' in this way, it is still no coincidence that silencing is systematically unjustly distributed in ways that adversely affect the experiences of those facing independent forms of subordination and oppression. This important fact, I think, is better brought out by using our terms in a way that does not obscure its substantive importance. Similarly, McGowan (2019) reserves the term for cases that are *systematic*, and again I agree that these are the most interesting cases, but hold that we can best build up to our understanding of systematicity by naming the thing that we take to be systematic.

silencing *locutionary* silencing, since it cuts off communication before speech. Locutionary silencing is the sense in which non-disclosure agreements silence, and it is this sense in which campus conservatives claim to be silenced by social censure. It is easy to understand why this kind of silencing would be unequally distributed so as to fall more heavily as a burden on those belonging to oppressed classes. After all, creating or resisting disincentives for someone to speak their mind is an exercise of power, and power is unequally distributed between the members of privileged and oppressed classes.

It is highly plausible that there is some element of locutionary silencing going on in the case of the available interpretation of the conversation in 'Baby It's Cold Outside'. This is because gendered norms of polite communication impose costs on being direct—especially with respect to communicating rejection. There are predictable sociological reasons why the female voice does not just say, 'put the fucking drink away and bring me my coat, you rapey toad', even if that is what she is thinking, and even on the available interpretation of what is going on. But locutionary silencing does not *suffice* to explain what is going on in this interpretation of the dialogue, because the female voice is not completely locutionarily silenced—merely directed toward politer, less direct, ways of communicating. And these ways of communicating are still sufficiently direct that those who hear the available interpretation can imagine saying exactly these things and communicating successfully with an appropriately receptive friend. So something more is required in order to explain how someone could actually speak their mind clearly and out loud and still face an obstacle to successful communication.

We have come to learn about many such possible forms over the last three decades, beginning with Langton's and Hornsby's treatments of *illocutionary disablement*, in which someone comes to lack the conditions necessary to perform illocutionary acts such as asserting or informing, and including Samia Hesni's (2018) recent account of *illocutionary frustration*, in which someone does have but is perceived by another as lacking the standing required to perform an illocutionary act, Mary Kate McGowan's (2014) conception of *sincerity silencing*, in which someone is interpreted as not being genuine sincere, and many others. Each of these theories provides a *model* for what *might* happen, which explains why, if that *were* to happen, there would be a formidable obstacle to communication. In general, these models explore simple possibilities left open by our best understanding of what is actually required for successful communication, and so the interesting questions about applying these theories largely concern the extent to which they capture what is going on in the real-world cases of silencing that we can identify, as well as in others that remain more controversial.

These—and other—models for how silencing can work differ not only over the details of what the mechanism of silencing is but also over the intelligibility of non-veridical experiences of silencing, the extent to which silencing of this kind can come in degrees, the extent to which this kind of silencing can be imposed by someone else, the degree of fault required of conversational interlocutors in order for this kind of silencing to obtain, and the ease with which the empirical conditions of this form of silencing can obtain. The richness of these differences suits different models of silencing for capturing different phenomena—and different layers of the same phenomena—which would otherwise be run together. So it is in that sense worth letting a thousand flowers bloom, the better to discover which bear which fruit.[3] The remainder of this chapter is offered in this same spirit.

2. Attributive Silencing

As its name suggests, attributive silencing is made possible because only some of our behavior, and not all, is properly attributable to us in the fullest sense. When I say that some behavior is attributable to you in the fullest sense, I mean that it reflects your self, rather than being simply attributable to your environment or to other causes working within you. For example, if you snap at me in part because you are hangry, your tone of voice may be attributable more to the fact that your digestive tract expects calories than to you. It is part of your behavior, but it is also properly overlooked in order to engage with you.

If you snap at me out of hanger, in this way, then if I understand this, it does not make sense for me to become defensive or to snap back. If I do get defensive about your tone, then it should be enough to defuse the situation for a common friend to point out that you have simply missed lunch, and the appropriate response should instead be to pass you a Snickers bar. Of course, when you snap out of hanger, the content of what you say may reflect your choice and judgment even while the tone in which you say it reflects your hunger and your circumstances. So judging which aspects of someone's

---

[3] Here again I follow and am inspired by McGowan's ecumenicism.

behavior merit responses and which are best overlooked as deriving from their circumstances requires care. But we do it all of the time.

When I first met my now wife's family, I was introduced to her grandfather, who had recently suffered a stroke. As he explained it to me, "last year I got stoked, and since then, everything is a little *whoo-hoo*." (You have to imagine the twinkle in his eye and a hand gesture showing that things go over his head.) At first, I saw no evidence of impairment—he was wily and funny and participated in every part of the conversation—but at some point I began to have trouble keeping track of the reference of his pronouns. He would say things like 'she is going to graduate school', when I was the only one that I knew of heading to graduate school, and 'he has gotten so tall', when my then-girlfriend was the only obvious candidate for who had gotten tall. It eventually became clear that the genders of the pronouns that he was using were coming out more or less random. At first this made it difficult to interpret him, but after a few hours I was able to simply overlook it. Instead of treating the gender of his pronouns as a matter of choice, I began to treat it as something to be ignored, as monolingual English speakers ignore tone unless trying to mimic speech, and as we ignore accents, once we get used to them.

So we are constantly interpreting one another by selecting out which aspects of their behavior are the signal that we must pay attention to, in order to engage with them as persons, and which are noise that must be overlooked. If we engage with the noise, then we are liable to deep misunderstandings of one another. For example, if I get defensive about the tone in your voice or give you a mini-lecture about not speaking rudely, I am missing the point and you are liable to escalate an unnecessary conflict between us. And if I correct my wife's grandfather by explaining that my preferred pronouns are 'he' and 'him', all that I am doing is picking a pointless fight. To avoid misunderstandings like this, we need to be able to distinguish the protagonist from her predicament. As Strawson (1962) famously observed, anger, gratitude, resentment, and admiration are all appropriate attitudes to have toward the aspects of someone's behavior or psychology that represent their agency as a protagonist, but make little sense as attitudes towards those that are merely consequences of their predicament.

But to interpret someone in this way requires constantly making interpretive choices—whether we think of them as choices or not. We must take a stand on which aspects of someone's behavior or psychology are the ones to be engaged with, and which are not. But as any physician will remind us, no way of determining something like this can be perfect. Any test that we use will be subject to false positives and to false negatives. If we identify some part of someone's behavior as attributable to them and it really is, then that is a true positive, which is great, and is hopefully the normal case—it lets us engage with them. And if we identify some part of someone's behavior as not attributable to them and it really is not, then that is a true negative, which is also great, and allows us to overlook hangry tones of voice and stroke-scrambled pronoun genders. But if we identify some part of someone's behavior as attributable to them and it is not, that is a false positive for attributability, and, as we've seen, that creates problems, as in the case where I pick a fight with my wife's grandfather over my gender.

But in addition to true positives, true negatives, and false positives for attributability, we are also bound to end up with false negatives for attributability. Indeed, no matter how well intentioned we are, the only way to avoid at least some false positives is to accept some risk of false negatives. In a case of a false negative for attributability, we identify some part of someone's behavior as not attributable to them when in fact it is. For example, you raise your voice in order to make clear to me how strongly you feel about something, but I write it off as hanger. Or you laugh out of genuine amusement at my jokes but out of insecurity I interpret this as simply baseline courtesy.

Whenever there is a false negative for attributability in this way, some important exercise of my active agency is overlooked or dismissed, as if it were merely the product of hunger, nerves, hormones, or alcohol. This is not by itself silencing, since of course not all behavior is an attempt to communicate. But since speech is a *kind* of behavior, the same principles that apply to behavior in general also apply to speech. When someone's communicative act is recognized by someone else but interpreted as not attributable to them when it really is, we have a false negative for attributability of a communicative act. When this happens, you are trying to communicate to someone, and they even recognize that you are trying to communicate with them. But they dismiss your communicative act as best overlooked and not engaged with, in the same way that it is right to overlook your hangry tone of voice or my wife's grandfather's pronoun genders.

This is a very distinctive way for your communicative act to be unsuccessful. It is not unsuccessful because you did not say anything, or because circumstances prevented you from performing your intended illocutionary act. Nor is it unsuccessful because your interlocutors mistook the situation for one in which you could not perform that illocutionary act, or mistook your intentions as not really intending to perform it. Rather, it is unsuccessful because even after recognizing exactly what illocutionary act you have successfully performed, your interlocutors interpret it as not mattering or being worth engaging with, because you are only saying this because you are hungry, or tired, or drunk, or being affected by drugs or hormones.[4]

This does not quite satisfy our definition for being silenced, because we said that being silenced requires you to face some *obstacle* to successful communication. But, strictly speaking, someone can misinterpret you without there being any obstacle to them interpreting you correctly. So we shouldn't quite say that false negatives for the attributability of a communicative act are themselves automatically cases of silencing. But typically, of course, it will not be random that someone misinterprets you in this way—something about the situation will make it more difficult for them to avoid this kind of false negative for attributability to you. And whatever makes that more difficult will therefore be an obstacle to you communicating what you would like to communicate. So in such cases it makes sense to say that you are *silenced*. I will descriptively dub this kind of silencing *attributive* silencing, in recognition of the role of the concept of attributability helping us to understand its nature.

Mary Kate McGowan introduced the concept of attributive silencing in chapter 6 of her 2019 book *Just Words*, and in an article version of the chapter, McGowan (2017). And in correspondence she credits it to a suggestion by Lauren Ashwell at an APA meeting. In the taxonomy of McGowan's book, it is *type-4* silencing; she now calls it *deep self* silencing. But McGowan only spends a couple of paragraphs on attributive silencing in these places. So there is still added value, I hope, in jumping up and down and pointing at it, as I have been endeavoring to do in this chapter. In the remainder, I will try to argue that attributive silencing may in fact be both *important* and *pervasive*.

3. The Shape of the Thing

Like all models for how silencing can occur, it is one thing to say what it would be for someone to be attributively silenced and for us to agree that

---

[4] It is therefore a kind of *perlocutionary* failure, in the classificatory terms of Langton (1993). Langton writes off perlocutionary failures as less interesting forms of silencing, in part because of her specific dialectical goals and the fact that legal freedoms of speech arguably do not include perlocutionary freedoms. Yet attributive silencing is not just any failure to achieve one's perlocutionary goals. It really is, for all communicative and interpersonal purposes, as if one had not spoken at all.

that is in principle a possible situation that someone could get into, and another to conclude that it is easy enough to get into this situation that it could be a common enough experience to be worth talking about. And it is yet another thing still to establish that the empirical conditions of this kind of silencing actually obtain in any of the original range of cases that motivated our inquiry.

One very important worry about attributive silencing derives from the thought that respect for persons always requires respecting their own authority to the effect that some particular aspect of their own behavior or character is in fact attributable to them. So while it may be *possible* to mistake the tone in my voice as coming from the fact that I am hangry, once I tell you, "and I'm not just snapping at you because I'm hangry, either—the way that you've been treating me has been bothering me for months," it would be at worst contemptuous and at best disrespectful for me to continue to interpret you as requiring no response other than a Snickers bar. So as long as we are able to communicate with one another, we can always, you might think, overcome any such interpretive obstacles.

But wait. *So long as we are able to communicate with one another*, this is of course true. But of course attributive silencing is itself an obstacle to communication. So if the respect in which someone is attributively silenced itself includes their very attempts to communicate that they are being attributively silenced, then they will not be able to communicate their way out of it. So the important question is: how realistic is this?

Take the case of Odysseus, who has heard from Circe that the song of the sirens is both bewitching and beautiful. We all know the story: he tells his sailors to tie him to the mast and to plug their ears with wax, so that they cannot hear the sirens and will not be tempted. But even more important to the story, of course, is that the wax will prevent them from hearing his own revised orders, once he has fallen under the sirens' sway. "Keep the wax in your ears," he warns his sailors, "else you will hear me issuing bad orders! None of those orders will be worth a whit, because I won't be myself—I will be under the influence of the sirens."

In the normal version of the story everything goes as planned. But it will be helpful for us to consider a different version of the story—one in which the wax slips from one of the sailor's ears. In my version of the story, luck would have it that the wax falls out just as the sirens are pausing in their song to take a breath. So instead of hearing their alluring song, he hears only the yells of Odysseus over the waves. Odysseus says, "thank the gods, man, that the wax has fallen from your ears. Come untie me from this mast so that we can pilot the boat safely. Earlier, when I asked you to tie me to the mast, I was under the influence of that vile witch, Circe. But only now has the song of the sirens cleared my head and have the scales fallen from my eyes. This is now me speaking clearly and in my own right. Come untie me!"

The sailor must make a choice between which Odysseus to believe. The earlier Odysseus, who claims to be in his own right mind and says that the words and actions of the later Odysseus do not speak for him and should be disregarded, or the later Odysseus, who claims to be in his own right mind and says that the words and actions of the earlier Odysseus do not speak for him and should be disregarded. In the original version of the story, context makes clear which of these is the right interpretation. But we can't always be sure while it is happening, and either answer refutes the naïve view that we must always defer to the privilege of an agent's own judgment. Fate can sometimes put us at odds with someone's own self-interpretation, and lead us to think—sometimes rightly, but also sometimes rationally even when we are not right—that they are wrong about themselves, and that they are only interpreting themselves in that way because they are so far gone.

This is exactly how we treat the alcoholic's assurances that they really can stop after one more drink and they are not yet out of control: we think, the fact that they think so is just one more way in which alcohol has them in its grip. It is how we treat the rebellious teenager who thinks they will have no regrets engaging in some risky behavior: we think, the fact that they think so is just another manifestation of their need to distance themselves from their parents. And it is how the sailor must treat Odysseus, no matter which version of Odysseus he elects to trust. We may not always get things right when we overrule agents' own judgments about which of their actions speak for them and which do not, but the fact that we *sometimes* get them right in this way provides a rationalizing explanation for why we do it even in cases in which we get things wrong in so doing.

We have also seen that attributive silencing is possible without anyone being the one who is *doing* the silencing of someone else. It is a condition that someone can fall into when the circumstances elevate the risk of false negatives for attributability of some range of their behavior to them. Sticking to our holiday theme, in the movie *Elf*, Will Ferrell's eponymous character arrives in Manhattan from the North Pole, explaining that he knows Santa. When he arrives dressed as an elf at the department store toy department, he is assumed merely to be in character when he makes such remarks—a classic case, I take it, of illocutionary disablement, as circumstances render him unable to obtain uptake as making genuine assertions. But later when he arrives for dinner at his birth father's home and says the same things, everyone understands that he is making assertions, but they instead interpret these as deriving from a delusion. The circumstance simply makes it hard for his interlocutors to get him right. Such a circumstance could, of course, be the result of someone's deliberate action, and fodder for an excellent plot line for a better sort of film. But it need not be.

Similarly, nothing about attributive silencing requires the audience to be at fault in any way. Not only is it possible to get someone wrong by mistakenly taking some behavior of theirs—even a speech act—as not being truly attributively to them through no fault of your own, but we have actually seen that it is *inevitable* that each of us will sometimes do so. For accepting the risk of at least some false negatives is the inevitable price of interpreting one another in a way that avoids at least some false *positives* for attributability.

This doesn't mean, of course, that all attributive silencing is due only to the operation of rational principles of interpretation. We'll see in the next two sections, for example, that one of the driving circumstantial features that can create an obstacle to being correctly attributively interpreted can be pervasive distortions in social values, which are almost never innocuous and almost always reflect existing forms of social power. And our interpretation of one another can be biased and self-serving, especially when we do not really want to listen to what each other have to say. But it is possible, at bottom, even in the best cases.

Yet even though it is possible for someone to misinterpret us solely through rational good will, there are few things as demeaning as for someone to get us wrong in this way. When they do, they see perfectly well what we are saying, and write it off as irrelevant and not worth engaging with. This cuts deep.

The paradigm, for me, of the deep cut of attributive silencing is Kant's treatment of his correspondent Maria von Herbert, as drawn out in Rae Langton's (1992) beautiful and deep essay 'Duty and Desolation'. After a brief correspondence with von Herbert that challenges his own interpretations of what his own views entail about the moral significance of honesty, Kant bundles up his letters without replying to von Herbert's latest and passes them on to another correspondent, referring to von Herbert as an "ecstatical little lady" and a "sublimated fantasy". Her remarks are safely interpreted as the result of feminine hysteria, and as such they do not require engagement and can therefore be used for other purposes—as an illustrative example for another correspondent.

The aspects of our behavior that are attributable to us are those that belong to us as persons, and hence are those that require engagement in order to relate to us as persons. But the aspects of our behavior that are attributable instead to our environments or to other causes acting on us or in us belong to us instead as embodied things. They require no such engagement in order to relate to us as persons—indeed, they can be properly overlooked and disregarded. So when someone treats something that genuinely belongs to us as a person as if it instead belongs only to us as a thing, they are in effect treating *us* as a thing. This is what I want to suggest is deeply true about Langton's description of Kant's treatment of von Herbert:

> Herbert, now deranged, is no longer guilty. She is merely unfortunate. She is not responsible for what she does. She is the pitiful product of a poor upbringing. She is an item in the natural order, a ship wrecked on a reef. She is a thing.

> And, true to Kant's picture, it now becomes appropriate to use her as a means to his own ends. He bundles up her letters, private communications from 'a dear friend', letters that express thoughts, philosophical and personal, some of them profound... These are not thoughts, but symptoms. Kant is doing something with her as one does something with a tool: Herbert cannot share the end of his action.[5]

It is quite astonishing, I think, that Langton provides such a beautiful and accurate description of attributive silencing in a paper first published just one year before her seminal paper on illocutionary silencing. Yet it is clear, I think, that Maria von Herbert has not been illocutionarily disabled, in Langton's sense. True, Kant is not listening to her—Langton says that "it is hard to believe that he has read her second letter" (1993, 499). But the problem is clearly not that he does not understand the letter to involve assertions. Rather, what is so hurtful about it is precisely that he thinks it can be safely written off despite this.

Despite not naming attributive silencing or identifying it as a distinctive phenomenon, therefore, I think that Langton has given us our most compelling illustration of its power and depersonalizing consequences.

---

[5] Langton (1992, 500).

4. Cases, and Cases

I will now show that both how easy it is to become attributively silenced and how easy it is to escape it must turn on the correct theory of attributability, and hence, I will suggest, on the correct theory of persons. And so consequently, both how much attributive silencing we should expect to find, and the weight of the burden imposed on those who are attributively silenced, must depend on the correct theory of persons as well. As we will see, on my favored theory of persons we should expect to find a great deal, and the burden is often heavy indeed.

The problem is that since attributive silencing results from circumstances that make it more likely for others to land on false negatives for attributability for certain kinds of communicative acts, the question as to which sorts of circumstances will do so must be answered by the more general question as to what provides evidence as to what is attributable to someone and what is not. And in order to know what provides evidence as to what is attributable to someone and what is not, we need to know, in general, what it takes for something to be attributable to someone at all. That is, we need a general theory of the conditions of attributability.

Although he did not himself use the word 'attributability', Harry Frankfurt's (1971) 'Freedom of the Will and the Concept of a Person' offers a seminal treatment of the important distinction that came to be recognized as the distinction between what is attributable to someone and what is not. But it also, as exhibited by its title, advances the important idea that our theory of attributability should flow from our general account of persons. This idea is simple, and I think inescapable: since you are a person, the question of which of your behavior belongs to you underivatively is the same as the question of which of your behavior belongs to a *person* underivatively. But this question cannot be answered without knowing what makes you a person. And the question of which behavior belongs to you underivatively is just the question of what is attributable to you. Hence, our theory of attributability must flow from our general account of persons.

Frankfurt's own theory of attributability from this paper both illustrated this point perfectly and will serve for us as an illustration of how much the ease of silencing will depend on the true theory of interpretation. On Frankfurt's own view, a person is a creature with the capacity to determine their actions by their volitions. Hence, the actions that are genuinely attributable to you are those that are caused by your volitions. Frankfurt's is a picture on which you are, more or less, your will, and so behavior that is caused by your will is *ipso facto* directly caused by you, and behavior that is not caused by your will is yours only in a derivative way—by being the behavior of the body in which your will is located.

Now, if Frankfurt's theory is right, then the question of which of your behavior is genuinely attributable to you is going to be a question of the psychological causes of that behavior. If it is caused by volitions, then it is, and otherwise it is not. So only evidence of the lack of operation of volitions can be evidence that some behavior (including a speech act) is not attributable to you.

Now, given the right sort of background misinformation, anything could of course constitute evidence of this. For example, to pursue an example from the last section, if there is a widespread strong but mistaken belief that sane adults do not believe in Santa and hence do not have volitions to tell other adults about Santa, then it may possibly be difficult for someone with first-person acquaintance with Santa to explain this without their interlocutors wrongly taking this speech to not genuinely be attributable to them.

But it is not clear, I think, why we should expect this sort of misinterpretation of the operations of people's psychology to be prevalent or systematic. And, indeed, the more we learn about how human psychology works, it would seem, the better we should get at avoiding false negatives for attributability. So science and education, it would seem on Frankfurt's view, should do much to minimize the extent of attributive silencing.

But more: Frankfurt's account leaves *open* the possibility, explored in the last section, that someone could be attributively silenced both with respect to some first-order speech and simultaneously with respect to their attempts to communicate that their first-order speech is being attributively silenced. But aside from cases in which this is a simple byproduct of the fact that someone is interpreted as what Frankfurt calls a *wanton*—someone whose volitions are *never* effective and hence whose behavior is *never* attributable to them— there is no reason to think that this is an especially likely combination. The problem is that the effectiveness of volitions to communicate that one is being attributively silenced seems too independent from the effectiveness of volitions to communicate anything else. And since each of us should have better evidence of the causes at work in our own psychologies than of those at work in others, it provides an especially good reason to take someone's own testimony seriously, that some of their other speech really is genuinely attributable to them.

So if Frankfurt's is the correct account of attributability, then I think we should conclude that attributive silencing should be expected not to be particularly common, and that in most cases it should be relatively easy to escape, simply by telling our interlocutors that they are misunderstanding us and that our attempts to communicate really are attributable to us and so should not be disregarded. And it predicts that science and education should tend to diminish the extent and significance of attributive silencing.

Frankfurt's account of the conditions of attributability can be helpfully contrasted with my own. Whereas for Frankfurt attributability is a psychological condition that requires a special kind of cause, I have come to be deeply skeptical of accounts of attributability that amount in essence to locating the homunculus somewhere in your head whose actions count as truly yours. As Gary Watson (1975) pointed out long ago, what makes second-order desires so special? But the same question goes, I think, for *values*, Watson's own answer to the source of which actions are genuinely your own, and for any other psychological answer as well.

A different way to get into skepticism about a psychological answer to which behavior is attributable to you begins by comparing different causes that can affect your behavior. Sometimes non-rational internal causes like hormone levels or drugs can cause you to act in ways that are not yourself, it is true, but sometimes it is those very same causes that allow you to overcome obstacles to being truly yourself. And sometimes, I think, one and the same cause can do both at the same time—an observation many make about the effects of mood-altering drugs. Similarly, we know from situational psychology that sometimes external circumstances can prompt behavior in ways that we take to be more attributable to your environment than to you, but sometimes it is those same prompts, I suggest, that allow you to overcome obstacles to being truly yourself.

I therefore maintain that it is not the causes of our behavior—either internal or external—that determine whether it genuinely reflects us or not. Rather, it is something holistic about how all of these causes jumble together to determine the whole course of our lives. And that holistic thing, I suggest and have argued elsewhere, is that persons like you and me are *interpretive objects*. We are, each of us, constituted by the best interpretation of what counts as truly us and what does not.[6]

The project of deciding what is attributable to someone and what is not, therefore, requires looking at their life as a whole in order to decide what fits into it and what does not. When I interpret the snap in your voice as hanger

---

[6] Schroeder (2019); see also Schroeder (forthcoming).

speaking, I am deciding that an interpretation of you that leaves the snap out offers a better picture of how your life hangs together than an interpretation that leaves it in. And when Kant decides that Maria von Herbert's meditations on duty, the moral law, and honesty are the product of feminine hysteria rather than deep engagement with his arguments, he is deciding that this interpretation offers a better picture of how her life hangs together than the alternative.

Just one piece remains, in order to be able to draw out the contrasts for attributive silencing between my own account and Frankfurt's. And that is that I hold that our interpretations of one another are *evaluatively loaded*— not just in the sense that we are looking for the best interpretation of one another, but in the more specific sense that, other things being equal, our interpretations of one another are *biased towards the good*. That is, there is a role for something like a kind of *charity* in our interpretations of one another. This is something that I have argued for elsewhere—most prominently in Schroeder (2019)—and here I will simply rely on it in order to generate a point of stark contrast to Frankfurt, without being able to defend it at any greater length.

Since the question of which of your behavior is genuinely yours and which is not is an interpretive question, there is much more room for disagreement about its answer than there is about the psychological question of the causal etiology of behavior in second-order desires. And it is a straightforward corollary of the general fact that there is room for disagreement in interpretation that it can easily happen that my interpretation of you is different from your interpretation of you. But whenever my interpretation of you differs from your interpretation of yourself by attributing *less* to you than you attribute to yourself, you are going to have an experience as of being attributively silenced. That is because, *by your lights*, I have hit on a false negative for attributability.

This explanation of why you can so easily have an experience as of being attributively silenced does not, of course, tell us whether you are actually attributively silenced, because it is independent of which of our interpretations is the correct one. But given that it is so easy to disagree in our interpretations of you, it must be almost as easy for us to disagree but you be right. And so, as long as there is some feature that explains why I am interpreting you wrongly to count as the *obstacle* to your successfully communicating with me, it turns out that attributive silencing can actually happen quite easily, if my account of attributability is on the right track.

Moreover, because of the role of values in interpretation on my account of attributability, my account should also lead us to expect that situations in which you are attributively silenced with respect to your efforts to communicate that you are being attributively silenced are much more likely to be the norm than the exception. That is because the role of values in interpretation makes it far from independent whether your telling me that the tone in your voice really speaks for you and not just the hanger belongs as part of an interpretation of you that is biased toward the good, and whether the tone in your voice itself does. If excluding that tone of voice from my interpretation of you helps me to bias it toward the good, then it is likely that excluding your protestations will do so as well.

Compare again the case of Odysseus. When he leaves the comfort of his discussions with Circe and sails out to hear the sirens' song, his self-interpretation shifts. On shore, talking to Circe, he anticipates being tied to the mast and wanting to sail toward the sirens and issuing orders to his crew to do so, and he disavows these desires and acts as not really him—being instead the effects on him of the sirens' song. Yet out at sea he sees things differently. Now he thinks of his planning on shore for how to prevent the sailors from hearing him as a mistake and not reflecting his true choice, but thinks of his own current desire to sail towards the sirens as authentic, and of his orders to his crew as truly his own.

What, then, changes in Odysseus, to flip this switch not only in his desires and his actions but also in his self-understanding? My answer is that it is a switch in his values. On the shore, he values safety and hearing beautiful songs, but not actually sailing toward the sirens. But once out at sea, sailing toward the sirens becomes something that he now sees as itself good, over and above listening to their song. This switch engenders a shift in which interpretation of himself is best biased towards the good. And that shift leaves him just as confident that his new protestations that he is now in his right mind and it is his earlier orders that should be disregarded are truly attributable to him, as he is that his desires and orders are.

5. Oppression and Systematicity

In the last section, I argued that exactly how much attributive silencing we should expect to observe and how difficult we should expect it to be to escape being attributively silenced is going to vary, according to which theory of the conditions of attributability we accept. If some account like

Frankfurt's is on the right track, I argued, then attributive silencing should be relatively uncommon, and it should in most cases be relatively easy to escape, lessening any burden that it imposes. And we should expect science and education to have the power to mitigate the scope and effects of attributive silencing over time.

But if some account more like mine is on the right track, and attributability is holistic, then we should expect attributive silencing to be much more common, and we should not be surprised if it is actually quite difficult to escape, because the very same conditions that silence you with respect to some topic also silence you with respect to your efforts to communicate what is genuinely attributable to you, as they do with Odysseus. And since science and science education do not by themselves promise to narrow the range of reasonable disagreements about values or interpretation, we should not expect science or education to mitigate the scope or effects of attributive silencing over time—on the contrary, we should expect it to be essentially ineliminable.

Now, I have not offered any arguments for my own account of attributability in this chapter. I have largely only used it to contrast with Frankfurt's, in order to illustrate my point that much about the expected scope and burden of attributive silencing will depend on our theories of attributability and hence of persons. Yet if my discussion of attributive silencing has led you to think that it sounds alien and rare, this itself should lead you to expect that my account of attributability is probably on the wrong track. And if instead you recognize the feeling of being dismissed as Kant dismisses Maria von Herbert and the concordant feeling of being ignored in a way that feels depersonalizing, like you have been reduced to a thing rather than a person, then you should expect, I think, even in the absence of other arguments, that something broadly like my account of attributability really is on the right track.

So let us return to 'Baby It's Cold Outside'. On the interpretation of the song that I have called the 'available' interpretation, the female voice is genuinely trying to communicate that she desires to leave, but somehow failing in some way. It could be, of course, that her failure is entirely the fault of her interlocutor, who understands perfectly well that she sincerely wants to go and has said so, but is pretending that they are instead in the flirty circumstance envisioned by what I have called the 'intended' interpretation of the song, either out of a sense of denial at being rejected or, even more perniciously, in an attempt to bully or gaslight her into submission. But philosophical accounts of silencing offer us tools for understanding what else might be afoot that could make it difficult for her to communicate with him.

All philosophical accounts of silencing divide into a combination of conceptual and empirical components. The conceptual component of an account of silencing identifies what it would take for a certain kind of silencing to obtain, explains why that condition obtaining would amount to something worth being called 'silencing', and explores the consequences of what would follow from this variety of silencing. Whereas the empirical component of an account of silencing attempts to establish that the conditions of this kind of silencing actually do obtain as a matter of empirical fact in precisely the sorts of situations that we have identified as involving some sort of silencing, and of which we therefore would like to give the kind of insightful explanation facilitated by philosophy.

It is a striking fact that the vast majority of philosophical accounts of silencing divide the work up between conceptual and empirical components in the same highly inequitable way. Nearly all such accounts make it *easy* to establish the conceptual burdens of their account of silencing, but carry considerable burdens when it comes to establishing that the conditions of their variety of silencing actually obtain in many ordinary conditions in which people do seem to be silenced—including, paradigmatically, conditions of romantic rejection and sexual refusal, as in the 'available' interpretation of 'Baby It's Cold Outside'.

For example and paradigmatically, on the account of Langton (1993), women's acts of sexual refusal are illocutionarily disabled because of the authoritative exercise of power by the creators of pornography. Langton's account is, I suggest, easy to defend on conceptual grounds, once she and Hornsby (1995) have pointed the way for the rest of us. It really is possible to be illocutionarily disabled from being able to perform particular kinds of speech act, and it really is possible for this to happen with respect to assertion, as illustrated by the example of *Elf* in the department store in Section 3. The main difficulty with applying Langton's analysis to the case of sexual refusals—even setting aside her pursuit of the dialectical goals of defending the 'free speech' argument against pornography—is that it is difficult to empirically establish that the creators of pornography have the right kind of power to perform this speech act, and to explain how, even if they did, it could exert an effect on the conversational dynamics even in exchanges with men who do not actually view pornography.

Attributive silencing as conceived of through the lens of my interpretive account of persons offers an illuminating contrast. Instead of dividing up the labor between conceptual and empirical components in such a way as to shore up the conceptual component but place the burdens on the empirical component, under my account the conceptual basis of attributive silencing is very demanding indeed. In order to even believe in the possibility of attributive silencing that works in the way that I claim it does, you have to go in for a highly substantive theory of the nature of persons. This is a very high conceptual burden.

Yet one of the payoffs of this quite substantial conceptual burden is that the empirical commitments required in order to apply attributive silencing are very slight indeed. All that is required in order for women to face a systematic obstacle to their communicative efforts, on the full package of views that I have set out in this chapter, is that there be widespread and systematic distortions of value in their community. Because values are an important input to personal interpretation, having systematically mistaken values is all that it takes for their interlocutors to end up with systematically wrong personal interpretations of them. So a widespread distortion in values is all that is required, in order for women to be systematically attributively silenced, if I am right about the conditions of attributability.

The question of whether there is such a widespread distortion in values is an empirical one, of course, but it is empirical in the way that 'most people have two arms' and 'air is lighter than water' are empirical. It is something that we all know and recognize and encounter throughout our lives in a gendered social world that has been shaped by the power of men and by the power of women whose exercise of power is unthreatening to men. So the empirical component of my account is just the price of entry to recognizing the existence of gender norms in the first place.

This doesn't mean that I am suggesting that the dynamics of romantic rejection and sexual refusal can only be explained by the mechanism of attributive silencing. On the contrary, as I already suggested earlier, I expect that there is a rich fabric of overlapping and interplaying phenomena at play in such cases. So I am not offering attributive silencing as a competitor to other accounts. Yet I am inclined to think that if you can come so far as to accept anything like my account of the conditions of attributability, then the conclusion that it systematically affects romantic rejection and sexual refusal must be virtually irresistible.

The problem is that in order to have such pernicious effects, the systematic distortions in gender values do not need to *disvalue* anything about women at all. All that is required in order to create this effect is that some kinds of thing are valued *more* from women than from men—for example, caring, nurturing, or sex. So long as women's caring, nurturing, or sex is valued more than men's, interpretations of women that are biased toward the good will also be biased—and hence distorted—towards offering greater roles for caring, nurturing, and sex. It will be harder for women to be understood as *really* declining sex, just as it will be harder for women to be understood as *really* being uninterested in listening to your problems, and harder for women to be understood as *really* suffering from postpartum depression. Though there may be additional features that complicate some of these cases and not the others, there is at least one layer on which these phenomena are alike.

Yet my account is not perfectly gentle to men's overoptimistic construals of their interlocutor's intent either. Though on my account it is *possible* to arrive at a false negative for attributability of someone's behavior or speech without any ill will, in actual fact this is particularly likely to happen when wishful thinking opens us up to pay special attention to evidence that would support the hypothesis that we hope will turn out to be true. And it is surely true that the pressures of wishful thinking would point systematically in the direction of granting less credit to the hypothesis that an act is a genuine rejection or genuine refusal. So if that is right, then it is highly likely that most instances of this kind of misinterpretation are at least partly self-serving—even if they are also smoothed by social circumstances.

Now, when Langton originally advanced her account of illocutionary disablement, it was in the service both of defending the intelligibility and truth of Catharine MacKinnon's broader claims that pornography silences women, and of defending that claim's role in the free speech argument against pornography. If I am right so far about the social conditions that contribute to the silencing of women's sexual refusals, then I think it is safe to conclude that most of what would conventionally be called 'pornography' does in fact play a role in the sustenance and promulgation of distorted gender values. And if I am right, then this effect is not limited at all to work that portrays women saying 'no' when they mean 'yes', to work that denigrates female pleasure, or to work that portrays male dominance in sexual relations. It is instead a pervasive feature of work that glamorizes and promotes women as sexual beings in a social context in which women's sexual contributions are valued differently than men's. In other words, I hope, if I am right, to have captured one element of truth in MacKinnon's claim that pornography silences women.

Yet if I have done so, it has been without offering any new means to validate the free speech argument. For the role of pornography in sustaining this set of values, on this view, does not appear to be any different from many other forms of media that glamorize women's sexuality. Indeed, it does not appear to be any different from the song 'Baby It's Cold Outside' itself, which, even on the intended interpretation, glamorizes both feminine coyness and ultimate assent, and, with these, the idea of masculine sexual pursuit. So if this is right, then there are likely no easy solutions. We have our work well cut out for us. But getting closer to the truth about what is genuinely of value—that is, making a bit of progress in normative ethics—is a start.[7]

References

Caponetto, Laura (2021). 'A Comprehensive Definition of Illocutionary Silencing.' *Topoi* 40(1): 191–202.

Dotson, Kristie (2011). 'Tracking Epistemic Violence, Tracking Practices of Silencing.' *Hypatia* 26(2): 236–57.

Frankfurt, Harry (1971). 'Freedom of the Will and the Concept of a Person.' *Journal of Philosophy* 68(1): 5–20.

Hesni, Samia (2018). 'Illocutionary Frustration.' *Mind* 127(4): 947–76.

Hornsby, Jennnifer (1995). 'Disempowered Speech.' *Philosophical Topics* 23(2): 127–47.

Langton, Rae (1992). 'Duty and Desolation.' *Philosophy* 67(262): 481–505.

Langton, Rae (1993). 'Speech Acts and Unspeakable Acts.' *Philosophy and Public Affairs* 22(4): 293–330.

McGowan, Mary Kate (2014). 'Sincerity Silencing.' *Hypatia* 29(2): 458–73.

McGowan, Mary Kate (2017). 'On Multiple Types of Silencing.' In Mari Mikkola, ed., *Beyond Speech: Pornography and Analytic Feminist Philosophy.* New York: Oxford University Press, 39–58.

McGowan, Mary Kate (2019). *Just Words.* Oxford: Oxford University Press.

Maitra, Ishani (2009). 'Silencing Speech.' *Canadian Journal of Philosophy* 39(2): 309–38.

Mason, Elinor (ms). 'Sexual Refusal: The Fragility of Women's Authority.' Manuscript.

---

[7] Special thanks to Mary Kate McGowan, Gwen Bradford, Jan Dowell, David Shoemaker, and to the participants in my fall 2019 graduate seminar on attributive silencing at the University of Southern California.

Schroeder, Mark (2019). 'Persons as Things.' *Oxford Studies in Normative Ethics* 9: 95–115.

Schroeder, Mark (forthcoming). 'Treating Like a Child.' Forthcoming in *Analytic Philosophy*.

Strawson, Peter (1962). 'Freedom and Resentment.' *Proceedings of the British Academy* 48: 1–25.

Watson, Gary (1975). 'Free Agency.' *The Journal of Philosophy* 72(8): 205–20.

9

The Standing to Forgive

Maria Seim

1. Introduction

Consider this example: Your close friend, A, has been emotionally abused by their partner, B, for years without you knowing about it. One day A finally breaks off the relationship with B and seeks your support in the process. Your reactions towards your friend's partner might take many forms: resentment, indignation, anger, disappointment, etc. All of these reactions are examples of reactive attitudes we can call *blaming emotions*.[1] Simply put, your emotional reactions manifest that you blame B for what they have done to A. Over time, the abusive partner might change their ways. They might realize that what they did was wrong and that they have hurt A badly and ruined the relationship; they might change their behavior, go to therapy, and finally ask for A's forgiveness. A might offer B forgiveness, or they might not. Personally, you might continue to blame B for what they did to A independently of whether A has forgiven B or not. It is less likely that B will ask for *your* forgiveness, even though you also blame them for what they did to A.

The philosophical literature on the topic of third-party forgiveness testifies to our diverging intuitions about forgiveness in cases like this. Some see it as obvious that third parties can forgive, while others do not. Third parties blame wrongdoers for what they do to others—regardless of whether they are close relations or complete strangers. Third parties are often emotionally involved in wrongdoing that is not directed at them, and (absent special circumstances) their blame can be both fitting and appropriate.[2] If forgiveness

---

[1] I use the terms 'resentment' and 'blame' interchangeably in this chapter. I consider resentment to be the most common, but not the only, blaming emotion. Because 'resentment' is the term most frequently used to talk about blaming emotions in the forgiveness literature, I continue using the term here.

[2] I am here, and in the rest of the chapter, borrowing the terms 'fittingness' and 'propriety' from D'Arms and Jacobson (2000). D'Arms and Jacobson use these terms to explain how emotions can be fitting but not appropriate. On their account an emotional response is fitting if it correctly represents its object, but this is only a pro tanto reason to express the emotion. Other

is the best indicator that someone has ceased to blame, it intuitively seems right that third parties should be able to forgive wrongdoers as well. This intuition is also manifested in everyday speech. Third parties often express their ability or inability to forgive someone for what they have done to someone other than themselves: "I will never forgive B for what they did to A", or "I have finally forgiven B for what they did to A". Further, we can identify several moral reasons in favor of third-party forgiveness. We seem to think that it can be important for the third party to get over their blaming emotions, and for the repentant wrongdoer to be forgiven so they can go on with their life. Further, it might be important for the victim that third parties also forgive those who have wronged them. Lastly, in situations where the victim is dead, or unreasonable—that is, refuses to forgive even though we think they ought to—we seem to think it is important for the wrongdoer to be forgiven by third parties.

Nonetheless, the majority view in the philosophical literature argues that only victims have the standing to forgive.[3] Third-party forgiveness can be seen as disrespecting the victim's right to decide whether or not to forgive their perpetrator (Govier and Verwoerd 2002: 101). Further, the emotional response the victim has towards the wrongdoer is *her* emotional response, and she is the only one that can overcome it (Govier and Verwoerd 2002: 101). Recent contributions to the debate have, however, challenged the majority view and presented arguments in favor of third-party forgiveness (MacLachlan 2008 and 2017, Pettigrove 2009, Chaplin 2019). This chapter addresses these recent contributions and argues to the contrary that third-party forgiveness is impossible, because the very nature of forgiveness makes it the prerogative of the victim. In other words, only the victim has the standing to forgive. As will become clear, this entails that some cases of putative forgiveness are, in fact, not forgiveness, but something else.

The chapter proceeds as follows. In Section 2 I clarify the terms 'victim' and 'standing'. In Section 3 I present the recent arguments in favor of third-party forgiveness and divide them into two camps: one deals with the nature of forgiveness and attempts to show that the nature of forgiveness is

---

considerations could make the response all things considered inappropriate—despite it being fitting. In this context I use the terms to show how fittingness and moral or prudential propriety can come apart in relation to blame and forgiveness. Blame and forgiveness are fitting if the wrongdoer is blameworthy or forgiveness-worthy, but this only gives us a pro tanto reason to blame or forgive: there might be other considerations that make blame or forgiveness inappropriate.

[3] See Murphy and Hampton (1988), Haber (1991), Govier and Verwoerd (2002), Owens (2012), Zaragoza (2012), Walker (2013), and Warmke (2016).

compatible with third-party standing to forgive, while the other argues for third-party standing to forgive based on the moral reasons we have in favor of it. I discard the arguments for the moral propriety of third-party forgiveness because they are secondary to arguments based on the nature of forgiveness. In Section 4 I argue that what has become known as *the standard account of forgiveness* does not exclude the possibility of third-party forgiveness. The standard account of forgiveness tells us that forgiveness is the overcoming of resentment for the right reasons. When the wrongdoer repents, the victim has a reason of the right kind to overcome her blaming emotions. Importantly, however, I argue that this is not enough to explain what forgiveness is. The standard account is not able to explain how forgiveness can be voluntary. In Section 5 I develop a new account of forgiveness, where overcoming resentment is not a necessary condition, and where forgiveness is a choice. On this account third parties lack the standing to forgive, and forgiveness is consequently impossible for third parties. Finally, in Section 6 I argue that denying that third parties have the standing to forgive does not leave us without resources to explain how wrongdoers can get closure if the victim cannot, or refuses to, forgive.

2. Clarifications

2.1 Victims

Trudy Govier and Wilhelm Verwoerd (2002) have suggested that we can better understand the problem of third-party forgiveness by expanding the victim pool. As mentioned above, third parties are emotionally affected by harms directed at close relations. This could in effect make them secondary or tertiary victims (Govier and Verwoerd 2002: 103). If this is the case, third parties will have the standing to forgive in virtue of being indirect victims of the wrongdoing.

This is an important point, and it seems obviously true that there is such a thing as indirect victims, and that these victims also have the standing to forgive. As several philosophers have pointed out, however, this does not solve the problem of third-party forgiveness (Pettigrove 2009: 585, Walker 2013: 163, Chaplin 2019: 8–9). We can conceive of the example given at the beginning of this chapter as a case where a third party blames B for what they have done to A without being a victim themselves. I accept that third parties can be indirect victims, but sometimes they are not victims at all.

More importantly, third-party forgiveness is not a scenario in which we are forgiving B for what they have done to *us*—we are forgiving (or not forgiving) B for what they have done to A.

Third parties are also not forgiving 'on behalf of' the victim. As Pettigrove puts it, "Forgiving B for a wrong he has done to C is not the same as forgiving B *for* C. In other words, it is not the same as speaking, acting, or feeling on C's behalf or in C's stead" (2009: 591). We can conclude, then, that when a third party forgives, they are not preempting the primary victim's forgiveness. Rather, they either forgive the wrongdoer for the damage they have themselves suffered as a result of the suffering of the primary victim, in which case they have standing in virtue of being secondary victims, or they forgive in addition to, or despite, the victim's forgiveness or lack thereof, in which case they might not have the standing to do so.[4]

Expanding the victim pool might highlight other consequences of wrongdoing and those affected by it, but it does not solve the problem of whether third parties (that are not themselves victims) have the standing to forgive.

2.2 Standing

The literature on the standing to blame and forgive is currently growing, but there is still no agreement on exactly what we mean by 'standing'. In the literature on the standing to blame we can identify (at least) two meanings of the word: standing either as a *right* (Fritz and Miller 2019) or as *authority* (Sabini and Silver 1982).[5] In the case of forgiveness I suggest we understand these two senses of 'standing' like this: 1) Standing as right can be understood as something that makes forgiveness *morally inappropriate* if one lacks it. If you lack the right to forgive someone, it would be morally objectionable if you still did so, but it would not be practically impossible. Imagine I was to forgive my friend's abuser because they were offering me a big amount of money. This would likely be a morally inappropriate kind of forgiveness, but the moral inappropriateness, or the lack of right to perform the action, does not by itself make forgiveness impossible. 2) Standing understood as special authority, on the other hand, makes forgiveness either impossible or

---

[4] It also follows from this that, in the case of multiple primary victims, they will all have standing. Thanks to an anonymous referee for making me aware of this.

[5] Per Milam makes a similar distinction in "How is Self-Forgiveness Possible?" (2017). For an overview of the different positions on the standing to blame see Fritz and Miller (2018), or King (2019).

pointless if one lacks it.⁶ One way of understanding this is in analogy to the doctrine of standing in law (Sabini and Silver 1982, Murphy and Hampton 1988: 21). If one does not have the standing to file suit, one simply cannot file suit. This way of understanding standing can also be compared to the idea of special authority in speech acts. Some speech acts, it seems, can only be performed by those who have requisite authority. The speech acts in question are illocutionary acts called declaratives and commissives (Searle 1979). Declarative acts are of the kind where uttering something changes the world in some sense: "one brings a state of affairs into existence by declaring it to exist" (Searle 1979: 13). Some examples of declarative acts are naming, marrying, and appointing (Pettigrove 2004: 371). Commissive acts are acts where the speaker commits to some course of action, like promising, guaranteeing, and betting (Pettigrove 2004: 371). Searle and Vanderveken (1985) argue that these speech acts have preparatory conditions that secure the felicity of the act. This doesn't mean, of course, that I cannot pretend, or attempt, to declare something without requisite authority. It just means that what I am doing is in fact not declaring. On this understanding of standing as authority, forgiveness will be conceptually inappropriate if one lacks standing. By 'conceptually inappropriate' I mean that what we are doing when we try to forgive without standing is not best described by the concept 'forgiveness'. In other words, if forgiveness is conceptually inappropriate under the given circumstances (that is, the putative forgiver lacks authority), what we are doing is not forgiving, but something else. Others refer to this as the *logical* possibility of forgiveness (Norlock 2009: 18).

Both ways of understanding 'standing' (as right or as authority) can be applied to forgiveness, and if third parties have standing in the latter sense, they might still lack standing in the former sense. It might be that one has the requisite authority to forgive, but that this forgiveness is morally objectionable. If one lacks standing in the latter sense, one simply cannot forgive; it is thus less interesting to examine whether this failed forgiveness is morally appropriate. This is not to say that we cannot evaluate the permissibility of a

---

⁶ I focus here on the possibility that lack of standing (understood as authority) makes forgiveness impossible, and not on the possibility that it can make forgiveness pointless. It seems right that forgiveness coming from someone who is not well positioned to offer it will be pointless because it does not 'work'. If a stranger on the street tells me they forgive me for something I did to my sister, for example, I will not consider myself forgiven. But it seems to me that, if we understand standing as authority in the former sense (as something that makes forgiveness impossible), the same will be true. In other words, it seems like forgiveness is pointless because it is impossible. In relation to forgiveness, then, standing as authority is best understood as something that makes forgiveness impossible and conceptually inappropriate if one lacks it.

failed attempt to forgive; my point is simply that if third-party forgiveness is impossible, what third parties are doing when they think they are forgiving, or trying to forgive, must be something else, and that we then have to evaluate the permissibility of this other action.

3. The Moral and the Conceptual Appropriateness of Third-party Forgiveness

As mentioned, there are different ways of arguing for third parties' standing to forgive. Some focus on the conceptual possibility of third-party forgiveness, while others argue for the moral appropriateness of such forgiveness. I will argue that these two strategies overlap with the two ways of understanding 'standing' mentioned above. Those who argue for the moral appropriateness of third-party forgiveness understand 'standing' as a right to forgive, while those who focus on the conceptual possibility of third-party forgiveness understand it as special authority. In this and the next section I will look at the two strategies in turn, starting with arguments for the moral permissibility of third-party forgiveness.

3.1 Arguments for the Moral Appropriateness of Third-party Forgiveness

Alice MacLachlan (2008; 2017) and Rosalind Chaplin (2019) both defend third-party forgiveness. Denying that third parties have the standing to forgive, MacLachlan argues, is over-individualistic, and it fails to take into account the "important roles played by witnesses, bystanders, beneficiaries, and others who stand in solidarity to the primary victim and perpetrator" (2008: 7). MacLachlan and Chaplin both rely on an idea of a 'relationship of moral solidarity' as the foundation that makes third-party forgiveness permissible. If we stand in a relationship of moral solidarity with the victim, they argue, we have standing to forgive (MacLachlan 2008, Chaplin 2019).

The term 'moral solidarity' is first used by MacLachlan (2008), and by it she means to capture the sense in which we can identify with what the victim has suffered or is suffering from the wrongdoing in question. MacLachlan uses the example of intimate relationships of care, like family ties and friendships—but she also mentions relations formed through shared social and political identification: "... those who have been affected or have witnessed hate crimes, or bystanders to particular kinds of social exclusion and violence" (2008: 9). In other words, we can gain standing to forgive by having undergone a similar offence, or through some other means that makes us particularly well positioned to understand what the victim has suffered. This opens up the possibility of what Chaplin calls 'victim-relative' standing (2019: 73). MacLachlan grounds standing in the ability to empathize with the victim, while Chaplin is less restrictive and allows standing to non-empathizing parties who stand in close relationships with the victim (Chaplin 2019: 82).

It is not clear that the practice MacLachlan and Chaplin are considering the permissibility of is actually forgiveness. MacLachlan and Chaplin both hesitate to give a precise account of the nature of forgiveness. MacLachlan states that she is "prepared to remain relatively agnostic about the precise nature of forgiveness" (2017: 139). Chaplin, on the other hand, favors what she calls 'attitude-based' accounts of forgiveness, even though she does not "attempt to give conclusive arguments against any particular view" (2019: 74). The important element of their accounts is that third-party forgiveness, or the refusal to offer third-party forgiveness, offers support to the victim. These accounts are thus focusing their defense of third-party forgiveness on the moral goodness and moral permissibility of such forgiveness. I therefore interpret them as using standing in the first sense of the word: as a right to forgive. What these views in effect are arguing is that those who stand in a relation of moral solidarity to the victim can have the right to forgive.

The important role friends, witnesses, and bystanders play for a victim of wrongdoing is undeniable, and MacLachlan and Chaplin are right to point this out. Third parties often play important roles as co-blamers, and help victims realize that their blame is justified. This is particularly important in cases where there is an imbalance of power, and the blame of the victim might not be enough for the wrongdoer to acknowledge the wrongdoing. It is not thereby established that what friends and bystanders are doing when they do (or do not) cease to blame a wrongdoer is actually forgiving (or refusing to forgive). I argue that, to establish the moral permissibility or appropriateness of third-party forgiveness, one must first give an account of the nature of forgiveness. If there is nothing in the nature of forgiveness that restricts who has standing to perform it, we can then go on to examine whether or not such forgiveness is morally permissible.[7] Regardless of

---

[7] As explained in Section 2, I am not claiming that we need to give a full account of forgiveness, but that we need to know what conditions must be met for forgiveness to be conceptually possible.

whether we think that what third parties are doing is actually forgiving, they are clearly doing *something*, and we can discuss the moral permissibility of this something. But this does not mean that the reasons we might have in favor of this practice are actually reasons for third parties to *forgive*.

This distinction can be understood in the same way as I understand fittingness and propriety above (note 2). It can be morally appropriate to forgive someone in the sense that we are doing something morally good (or at least neutral) by forgiving. While it at the same time might not be fitting. As mentioned, there are moral reasons in favor of forgiving, but these considerations cannot explain how third-party forgiveness is possible.

3.2   Arguments for the Conceptual Possibility of Third-party Forgiveness

When arguing for the conceptual possibility of third-party forgiveness, it is common to identify what the extant literature has suggested as necessary and/or sufficient conditions for forgiveness, to see if these conditions can be met by third parties (Pettigrove 2009, Chaplin 2019). In what follows I will look at the most popular suggestion for a possible necessary and sufficient condition for forgiveness, to see whether or not it excludes the possibility of third-party standing to forgive.

What has emerged as a standard definition of forgiveness is the idea that forgiveness necessarily involves the *overcoming of resentment* (Murphy 2003, Darwall 2006, Griswold 2007).[8] Or what Brandon Warmke has named the Resentment Theory of Forgiveness:

(RTF):   Forgiveness is (or at least crucially implicates) the overcoming of resentment (2015: 490).

There are roughly two versions of the RTF, one conditional and one unconditional.[9] Both versions see overcoming resentment as a necessary condition for forgiveness, but some argue that overcoming resentment is

---

[8] Other necessary conditions have been suggested, such as repairing the relationship between wrongdoer and victim, and/or a changed evaluation of the wrongdoer. I will not discuss these suggestions here because I think that they can be either discarded or subsumed under 'overcoming resentment', but see Pettigrove (2009) for an overview of the suggested conditions.

[9] Proponents of what I call conditional forgiveness include Kolnai (1973–4), Murphy and Hampton (1988), Richards (1988), Swinburne (1989), Novitz (1998), Hieronymi (2001), and

necessary, but not sufficient, for forgiveness. Conditional versions of the standard account are characterized by the fact that they add further necessary conditions that depend on action from the wrongdoer, while unconditional accounts see forgiveness as something that is up to the victim alone.

I will focus on Pamela Hieronymi's version of a conditional account here, as this is (probably) the most influential version of it. Hieronymi argues that forgiveness must be active or, in her terms, *articulate*. This means that forgiveness must be the rational revision of a judgment that undermines resentment (2001: 535). This is an important desideratum for forgiveness, because if forgiveness is not responsive to reasons, it can be conflated with the non-rational manipulations of our blaming emotions, and such non-rational manipulation would separate forgiveness from our agency (2001: 535).

In addition, Hieronymi argues that forgiveness must be *uncompromising*. Forgiveness must be uncompromising so as not to conflate forgiveness with other ways of overcoming resentment, like justifications, excuses, or condonation. For an account of forgiveness to be uncompromising, Hieronymi argues, we cannot revise the following judgments: 1) the act in question is a serious moral wrong; 2) the wrongdoer is responsible and can be blamed; and 3) you do not deserve to be treated this way (2001: 530). If we revise 1 or 2, the wrongdoer can be excused, and forgiveness is not necessary. And if we revise 3, we condone the wrong. According to Hieronymi, the only change in judgment that rationally undermines resentment is the judgment that the wrong done persists as a present threat. This is connected to her account of wrongdoing as a threat to the rights and interests of the victim, and of resentment as a protest against this threat (Hieronymi 2001: 530). For resentment to be rationally undermined, the wrongdoer must take back or dissolve the threat implicit in the wrong action. When the wrongdoer has done this, we can revise the judgment that the wrongdoing persists as a present threat (Hieronymi 2001: 548). One way to understand this is that repentance removes resentment-worthiness (the wrongdoer no longer deserves to be the target of the emotional response), even if it does not remove blameworthiness understood as attribution of responsibility.

Hieronymi's influential account demonstrates how we can revise a judgment, and rationally undermine resentment, without giving up on the judgment of blameworthiness and conflate forgiveness with excuse or condonation. In addition, the revision of a judgment, on Hieronymi's account, seems to make forgiveness active, and not something that just happens to us, thus not conflating forgiveness with forgetting or being distracted.

The idea that overcoming resentment relies on some sign of repentance or apology from the wrongdoer is widespread in the literature, and repentance is often seen as the only right kind of reason to overcome resentment (Milam 2019). When the wrongdoer genuinely repents, the reason we have for resenting them is rationally undermined: we seemingly have no reason to resent the wrongdoer anymore.

Repentance is often thought of as a moral debt the wrongdoer has to pay (Nelkin 2013: 16–17). We are here not thinking of material compensation but, rather, of emotional compensation. What exactly is expected from the wrongdoer will vary on a case to case basis, but the most important feature of repentance, as I understand it, is that the wrongdoer somehow communicates to the victim that they acknowledge the moral wrongness of their action; that they no longer endorse the action, or the meaning of the action; and that this expression of guilt, remorse, or regret appears genuine.[10]

What's important for my purposes at this time is that, no matter what role overcoming resentment plays in one's account of the nature of forgiveness, overcoming resentment by revising the relevant judgment for the right reason is not the prerogative of the victim. Third parties can have the same reasons a victim has to overcome resentment. As both Pettigrove (2009: 590) and Chaplin (2019) rightly point out, third parties also respond to wrongdoing with feelings of resentment or indignation. Further, third parties might also overcome their resentment as a response to something the wrongdoer does. If overcoming resentment for the right reasons is all that is needed for forgiveness, third parties can surely forgive.

4. Resentment and Control

To include the element of repentance as a necessary condition for forgiveness I suggest this revised version of the RTF:

---

[10] For an extensive account of repentance, see Linda Radzik (2009), ch. 2, "Repaying Moral Debts". Retributive theories focus on the pain or punishment of the wrongdoer, while restitution theories focus on the victim.

(RTF*):   Forgiveness is (or at least crucially implicates) the overcoming of resentment for the right kind of reason (that is, the wrongdoer has repented).

The standard account of forgiveness so far construed is not limited to victims. There is no reason why third parties should not be able to overcome their resentment or indignation as a response to the wrongdoer's repentance. At a closer look it turns out, however, that the RTF* is flawed. Overcoming resentment should not only be subject to the kind of rational control Hieronymi focuses on; it should also be subject to a kind of voluntary control.

As discussed, forgiveness must be subject to rational control, because if it is not, it can be conflated with forgetting or being distracted, and this would make forgiveness fall far outside our rational agency. There are, however, two levels of control at work here. There is the sense of control that I have called 'rational control' above. This is the kind of control Hieronymi relies on, and it has to do with how our attitudes are judgment-sensitive and responsive to reasons. When we make evaluative judgments, we are exercising our rational control. As an example, if we stop feeling resentment because we have changed our evaluation of the wrongdoer based on their repentance, we are exercising rational control. But there is another narrower sense of control that can also be used to describe what we want from an account of forgiveness: what we can call 'voluntary control'. It could be argued that we not only want forgiveness to be reasons-responsive but also want to be able to decide, or choose to forgive *at will*.

On the account of forgiveness outlined above, the wrongdoer's sincere repentance could make it the case that the victim rationally ought to forgive them. Importantly, however, we do not want forgiveness to become a duty. In a case of severe wrongdoing where the wrongdoer has repented, we do not want to say that the victim is doing something morally wrong, or is being irrational, if she refuses to forgive. Forgiveness should be up to the victim in the sense that it depends on what reasons she has, but even if she has the right kind of reasons to forgive, she should be able to decide not to forgive.

The difficulty is to explain how forgiveness can be subject to rational ought-statements, and at the same time be something the victim can decide to do (or not to do) at will. The conditional version of the standard account of forgiveness is not able to explain how both things can be true at the same time. The rational revision of the judgment of resentment-worthiness (or persisting threat, on Hieronymi's account) does not seem to be voluntary in the right way. In other words, the rational revision of our judgments is not something we have voluntary control over.

To see this, we need to take a closer look at RTF*. Brandon Warmke has argued that the RTF, as portrayed by Hieronymi (or RTF*, as I call it), contains the following three elements: "(1) The rational revision of some judgment that makes continued resentment rationally unfounded; (2) the 'com[ing] to see that our resentment is unfounded'; and (3) the consequent 'disappear[ance]'of resentment in the well-functioning psyche" (2015: 496). If all these elements are at play, the victim must, on pain of irrationality, forgive. In fact, as Warmke points out, the account "imbues wrongdoers with the power to *rationally obligate* victims to forgive" (2015: 503). Because, when the wrongdoer has repented, the judgment that justifies resentment must be revised and resentment is no longer rationally founded.

This challenges the idea that overcoming resentment is something we can do at will. Not to forgive, if the right reason is present, then, would be irrational. If you are not irrational, you will automatically forgive if the right reason to overcome resentment is present. This does not seem right, because it would remove the victim's ability to decide when she thinks the wrong-doer has repented sufficiently. This does not track the idea that forgiveness should be a choice in the narrower sense of voluntary control, and it implies that even when it is a response to the right reasons, overcoming resentment is not sufficient for forgiveness.[11] It looks like we need something more to be able to explain how forgiveness can be subject to voluntary control.

Furthermore, overcoming resentment might not even be necessary for forgiveness, because we can imagine cases where the victim forgives as a response to the right reasons while still feeling resentment. Imagine a monogamous relationship between A and B, where B is unfaithful to A. B is fully blameworthy and has no excuse or justification. B admits to the wrongdoing immediately and apologizes sincerely to A. B loves A, and they commit to do all in their power to make up for it. Many might agree that A has no duty to forgive B, but after a while A might still want to do so. It is very possible, however, that A continues to feel resentment towards B for what happened, at least occasionally. It is also very possible that A will continue to feel this resentment on occasion forever. But, according to the RTF*, if B has repented, feeling resentment should at some point become

---

[11] I think that Hieronymi could easily solve this by saying that even if one is rationally obliged to forgive, one can choose not to do so. I will spell this out in more detail below.

irrational for A. Further, on RTF* this continued resentment is evidence of the fact that A never really forgave B.[12]

I agree with Hieronymi (2001: 535) that our emotions are sensitive to reasons, and that we therefore have rational control over our attitudes. Nonetheless, we do not have the power to change our emotions directly (and sometimes not even indirectly), and at will. It is common to continue to feel angry, even though we come to realize that the object of our anger is not anger-worthy. If forgiveness depends on us ceasing to feel any blaming emotions towards someone who has wronged us, we might never be able to forgive, even though we have the right reasons and the will to do so. I consider this a problem for the RTF*.[13]

Overcoming resentment as a necessary condition for forgiveness turns out to push forgiveness out of reach of our intentional agency. On the one hand, overcoming resentment does not seem sufficient, because we can imagine cases where the wrongdoer has repented and thus rationally obligated a victim to forgive, even though they do not want to forgive. While, on the other hand, overcoming resentment does not seem necessary, because we can imagine cases where the victim chooses to forgive despite still resenting.

5. Absorbing the Cost

To solve the problem of how forgiveness can be subject to voluntary control, I argue, we must give up the standard account of forgiveness. Further, I will argue that what we do when we forgive is decide to *absorb the cost of the wrongdoing* (Hieronymi 2001: 551) and not hold it against the wrongdoer anymore (Nelkin 2013). Absorbing the cost and deciding not to hold it against the wrongdoer is sufficient for private, or unexpressed, forgiveness, but for the wrongdoer to actually be forgiven we must also express the attitudinal change and the decision that goes with it.[14]

---

[12] See Norlock (2009: 20) and Warmke and McKenna (2013: 196–7) for a similar point.

[13] It should be mentioned that if we think of emotional responses as character traits, we might have voluntary control over them. This would, however, not mean that they are direct responses to repentance but traits that we can develop and change over time. I will leave this possibility for a future endeavor and thank an anonymous referee for the suggestion.

[14] It might follow from this account that forgiveness must be expressed to be genuine forgiveness. I do not argue for this view here, but I do not see it as a problem that forgiveness cannot come about unless it has been expressed in some way. If there is such a thing as private forgiveness, this forgiveness will not have the normative powers we normally think of forgiveness as having.

The account I present portrays forgiveness as twofold: it is a decision to change one's attitude, and a speech act that changes the normative landscape. Third parties have the same reason to forgive as the victim does; they can also go through the attitudinal change as a response to the wrongdoer's repentance, but they do not have the standing necessary to decide to absorb the cost of the wrongdoing, or to perform the speech act necessary for the wrongdoer to actually be forgiven. To get to this conclusion I must first establish what forgiveness is if it is not the overcoming of resentment.

One suggestion for an account that does not depend on the overcoming of resentment has been suggested by Brandon Warmke and Michael McKenna (2013). Warmke and McKenna argue that overcoming resentment is a common element of forgiveness, but that it is neither necessary nor sufficient. On this view, forgiveness is not necessarily connected to the emotional response involved in blame but is, rather, seen as a norm-governed move in a conversational model of moral responsibility. Warmke and McKenna develop a paradigm-based account of forgiveness that does not present necessary and sufficient conditions at all, and instead takes a paradigm example of forgiveness as explanatorily basic.[15]

Warmke develops his view further (2015, 2016), with a focus on the norm-changing aspect of forgiveness. On Warmke's account, forgiveness is, in part, a speech act that changes the normative landscape. More precisely, Warmke argues that paradigmatic cases of forgiveness are "communicative acts that possess, not only behabitive and commissive force (per Pettigrove), but also declarative force" (2016: 697). As mentioned in Section 2, declarative acts have the power to change reality, but only those with appropriate authority can perform felicitous declarative acts.

I think Warmke's account of the speech act element of forgiveness gets something important right. One must have standing for the speech act to be felicitous, and if we lack standing, we will not be forgiving when we utter the words "I forgive you". Without standing forgiveness is conceptually inappropriate, or simply impossible. By this I mean that we will not be rendering the wrongdoer forgiven if we utter these words without standing to do so.[16]

---

[15] The main idea behind this method of accounting for forgiveness is that derivative practices can be seen in comparison with the paradigm example, and the practices that are far away from the paradigm will be less likely to be captured by the concept 'forgiveness'.

[16] It should be mentioned that forgiveness might also be expressed through actions or gestures. I am open to the possibility that we can felicitously communicate forgiveness without words as long as the wrongdoer is able to recognize these actions or gestures as forgiveness. Thanks again to an anonymous referee for pointing this out to me.

The norm-altering element of forgiveness, as explained by Warmke, partly explains the relation between forgiveness and our practice of holding morally responsible through blame. One way to understand this is to think that, when we blame, we impose obligations on the wrongdoer for them to answer for their wrongdoing. As Dana Nelkin explains, we incur obligations all the time, and when we harm others, we incur the obligation to make restitution and to "somehow make up for or in some way address the wrong itself" (2013: 18). How much 'making up', or repentance, is necessary for forgiveness to come about is (at least to some extent)[17] up to the victim to decide. When the victim decides to forgive, she consequently frees the wrongdoer from the obligation to continue to make amends, and, in Warmke's terminology, changes the wrongdoer norms. In addition to this, Warmke argues, we also change the victim norms when we forgive, by giving up our right to blame the wrongdoer (2016: 698). After we have forgiven, we no longer have the right to express blame towards the wrongdoer. Nelkin argues similarly that forgiveness is (in part) constituted by this act of releasing the wrongdoer from their obligation (2013: 16).

Seeing forgiveness as a declarative act explains how forgiveness can change both the victim and the wrongdoer norms. Nonetheless, by giving up on resentment and repentance as necessary conditions for forgiveness, Warmke fails to explain the psychological motivation that allows the victim to engage in the norm-changing activity of forgiveness in the first place. The strength of the RTF* is exactly this: it explains how we can overcome resentment as a response to repentance, and thus how forgiveness differs from other responses to wrongdoing. Consequently, by giving up on these conditions Warmke has no means by which to explain how forgiveness differs from excusing, condoning, or forgetting in an informative way. The account explains how forgiveness, when uttered, can change both the victim norms and the wrongdoer norms, but it does not explain what is needed for the victim to perform the speech act. As Govier and Verwoerd underline when talking about coerced forgiveness, "saying is not doing and one does not forgive by performing a speech act" (2002: 101). In other words, anyone can utter the words "I forgive you", but the utterance could be coerced or dishonest. I don't think Warmke is claiming that we, just by uttering "I forgive you", under any circumstances whatsoever have forgiven. The

---

[17] There are some restrictions on what the victim can expect here, and the repentance should be proportional to the deed, but only the victim knows exactly how much damage was done by the deed, so the exact amount of reparations will depend on them.

declarative act itself is not sufficient for us to surmise that forgiveness has actually taken place. The utterance must be a voluntary choice, and it must testify to some change of mind that has happened in the victim, but we need to specify what this psychological element of forgiveness—that allows for this speech act to be motivated by the right reason—is.

On the RTF*, the psychological process that would allow for the speech act to be felicitous is the overcoming of resentment. Further, overcoming resentment is made possible by the wrongdoer's repentance. On an account that does not take overcoming resentment to be necessary one might think that repentance becomes superfluous. I don't think that it is. Repentance came into the picture as the best candidate for a right kind of reason to overcome resentment, but it might be that repentance really is the right kind of reason for something else. I will suggest that repentance is the right reason for *absorbing the cost*. Absorbing the cost will here be seen as a voluntary action, as opposed to overcoming resentment, and as the psychological element of forgiveness which is needed for the norm-changing speech act to be felicitous.

Hieronymi points out that, even if the wrongdoer has repented, serious moral wrongdoing affects the victim in a way that repentance alone cannot repair (2001: 550). In cases of moral wrongdoing there will always be some amount of emotional or social damage done to the victim. As Margaret Urban Walker puts it, "While the wrongdoer's heartfelt retraction might dispel the threat implied by her treatment of the victim, she usually cannot repair all the damage, and in many grievous cases, she cannot repair any of the damage, her wrongful act has done to the victim. She can, so to speak, retract the threatening insult but not the injury" (Walker 2013: 500). It is thus not a damage of material or economic kind, or a damage to the moral worth of the victim. The damage in question will be there despite the wrongdoer's repentance, and despite the victim's overcoming of resentment. When the victim forgives, Hieronymi writes, she is agreeing "to bear in her own person the cost of the wrongdoing and to incorporate the injury into her own life without further protest and without demand for retribution" (Hieronymi 2001: 551).

Hieronymi does not explain in detail what is involved in 'absorbing the cost', but the way I understand it, absorbing the cost is a choice we have, and what we choose to do when we absorb the cost is to commit to change our attitude towards the wrongdoer and to not hold the wrongdoing against them anymore (by overtly blaming them).[18] This decision can come about as

---

[18] See Bennett (2018) for more on the idea that forgiveness involves a commitment.

a response to repentance, but for the wrongdoer to actually be forgiven it must also be expressed. It is thus both a speech act and a private and attitudinal phenomenon that says something about how we view the wrongdoer. We cannot decide not to feel resentment anymore, but we can decide not to express resentment and to commit to change our attitudes towards them.

Repentance is thus still necessary because it gives us a reason to decide to absorb the cost, and not because it gives us a reason to stop blaming, or to overcome resentment. Put in a different way, repentance can make someone forgiveness-worthy, rather that remove resentment-worthiness. Absorbing the cost is a voluntary action, and the victim decides when she thinks the wrongdoer has repented adequately for her to decide to change her attitude towards them. This does not mean that we cannot say that she ought to forgive after the wrongdoer has repented. As a third party we might judge the wrongdoer to be forgiveness-worthy, but we can neither rationally nor morally obligate the victim to make the decision to absorb the cost and perform the speech act of forgiveness.

This necessary feature of forgiveness is not something a third party can take it upon themselves to do. A third party does not have a cost to absorb; if they did, they would automatically become secondary victims, something that would then grant them the standing to forgive.

If we want to connect this to an account of the nature of resentment and blame, and understand resentment as blame (as many do), we will have to see private blame and expressed blame as separate. One can continue to feel blame/resentment after one has forgiven, but one cannot express this blame/resentment. In other words, getting rid of private blame is not necessary for forgiveness, but committing to overcome it by absorbing the cost of it, and deciding not to express it, is.

The reason why one cannot express it is, as Warmke argues, that when one forgives one gives up one's right to overtly blame the wrongdoer (2016). Absent special circumstances, anyone who is making a justified judgment of blameworthiness has a right to blame. As mentioned at the start, we normally do not object to third-party blame. This does not have to be spelled out in terms of rights, and one might just as well say that anyone can blame fittingly, absent special circumstances. For forgiveness to be fitting, the wrongdoer must express repentance in some way, and for forgiveness to come about, they must keep doing so until the victim has released them from the obligation to repent. When we perform the speech act of forgiveness, we release the wrongdoer from this obligation and give up our right to blame them; the wrongdoer is consequently free to stop repenting.

Giving up one's right to blame and release the wrongdoer from obligations is something only the victim can do. Only the victim can decide when the wrongdoer has repented and apologized sufficiently for them to decide to absorb the cost. And only the victim has the standing (understood as authority) to perform the speech act that changes the normative landscape and releases the wrongdoer from their obligation to repent. In other words, only the victim can perform the speech act felicitously. Importantly, it is only this speech act, performed by the victim, that can render the wrongdoer forgiven.

To conclude, the account of forgiveness I have presented has two elements. We first must decide to commit to change our attitude towards the wrongdoer (that is, absorb the cost), and not hold the wrong against the wrongdoer anymore. Second, we can perform the speech act that changes the operative norms and renders the wrongdoer forgiven. The first element is private and attitudinal, while the second is an action performed through a speech act. Only the victim has the standing to perform either, or both, of these elements. Only the victim has the standing to make the internal decision to absorb the cost as a response to the wrongdoer's repentance. This is the prerogative of the victim because only the victim can decide when the repentance is sufficient for them to no longer hold the wrongdoing against the wrongdoer. By making this decision the victim can go on to forgive actively through a speech act. Because only the victim can make this decision, only the victim has the authority to perform the speech act felicitously.

6. Implications of the View and Problem Cases

Some serious implications follow from accepting the account of forgiveness I have just presented. On this account forgiveness is dependent on the wrongdoer's repentance, and third parties do not have the standing to forgive. This means that many cases of putative forgiveness—such as third-party forgiveness and unconditional forgiveness—will have to be classified as something other than forgiveness. Many find this counterintuitive.

An obvious problem for my account is how to deal with situations where the victim is unreasonable and refuses to forgive even though they might have good reasons to do so, or where the victim is dead or otherwise unreachable. If third-party forgiveness, and unconditional forgiveness, are not possible, the wrongdoer cannot be forgiven if the victim refuses to do so, or if the victim is not alive or available. I acknowledge the importance forgiveness might have for the wrongdoer, and I am not claiming that the wrongdoer should be subjected to continued blame from third parties for all eternity. Importantly, however, I do not think that forgiveness is equivalent to overcoming resentment or ceasing to blame. Third parties can cease to blame, or "let go" of blame (Milam 2019), even if they cannot forgive.

Some of the alternatives to forgiveness might not be morally appropriate. Condonation, for example, might violate the victim's interests and rights. Similarly, forgetting might be wrong because we have a duty—to ourselves, and to the wrongdoer—to hold those who wrong us responsible. On the other hand, some alternatives might be almost as good as forgiveness. We might let go of blame because we realize the conflict we're in will not dissolve, and the wrongdoer will never apologize, so we simply have to agree to disagree. This would not be forgiveness, but it would allow us to resume relationships with people who have wronged us and let us let go of destructive grudges.

Lastly, my account of forgiveness excludes self-forgiveness, where self-forgiveness is understood as forgiving oneself for wrongdoing done towards someone other than oneself. My account is not incompatible with a person forgiving themselves for wrongdoing done towards themselves. On my account self-forgiveness would be better understood as letting go of self-blame.[19]

References

Bennett, C. 2003, "Personal and Redemptive Forgiveness," *European Journal of Philosophy*, 11, pp. 127–44.

Bennett, C. 2018, "The Alteration Thesis: Forgiveness as a Normative Power," *Philosophy & Public Affairs*, 46(2), pp. 207–33.

Chaplin, R. 2019, "Taking it Personally: Third-Party Forgiveness, Close Relationships, and the Standing to Forgive," in *Oxford Studies in Normative Ethics*, Vol. 9, pp. 73–94.

D'Arms, J., and Jacobson, D. 2000, "The Moralistic Fallacy: on the 'appropriateness' of emotions," *Philosophy and Phenomenological Research*, 61(1), pp. 65–90.

---

[19] For more on the possibility of self-forgiveness, see Milam (2017).

Darwall, S. 2006, *The Second Person Standpoint: Morality, Respect and Accountability*, Cambridge, Mass.: Harvard University Press.

Derrida, J. 2001, *On Cosmopolitanism and Forgiveness*, London: Routledge.

Downie, R. S. 1965, "Forgiveness," *Philosophical Quarterly*, 15:59, pp. 128–34.

Fritz, K. G., and D. Miller, 2018, "Hypocrisy and The Standing to Blame," *Pacific Philosophical Quarterly*, 99, pp. 118–39.

Fritz, K. G., and D. Miller, 2019, "The Unique Badness of Hypocritical Blame," *Ergo*, 6(19), pp. 545–569.

Garrard, E., and D. McNaughton, 2003, "In Defence of Unconditional Forgiveness," *Proceedings of the Aristotelian Society*, NS, vol. 103, pp. 39–60.

Govier, T., and W. Verwoerd, 2002, "Forgiveness: The Victim's Prerogative," *South African Journal of Philosophy*, 21(1), pp. 97–111.

Griswold, C. L. 2007, *Forgiveness: A Philosophical Exploration*, New York: Cambridge University Press.

Haber, J. G. 1991, *Forgiveness*, Savage: Rowman and Littlefield Publishers.

Hieronymi, P. 2001, "Articulating and Uncompromising Forgiveness," *Philosophy and Phenomenological Research*, 62(3), pp. 529–55.

Holmgren, M. R. 1993, "Forgiveness and the Intrinsic Value of Persons," *American Philosophical Quarterly*, 30:4, pp. 341–52.

King, M. 2019, "Skepticism About the Standing to Blame," in D. Shoemaker, ed., *Oxford Studies in Agency and Responsibility*, 6, pp. 265–88.

Kolnai, A. 1973–4, "Forgiveness," *Proceedings of the Aristotelian Society*, NS, Vol. 74, pp. 91–106.

MacLachlan, A. 2008, "Forgiveness and Moral Solidarity," in S. Bloch-Shulman and D. White, eds., *Forgiveness: Probing the Boundaries*, Oxford: Inter-Disciplinary Press.

MacLachlan, A. 2017, "In Defense of Third-Party Forgiveness," in K. J: Norlock, ed., *The Moral Psychology of Forgiveness*, Lanham, Md.: Rowman & Littlefield International, pp. 135–59.

Milam, P. 2017, "How is Self-Forgiveness Possible?" *Pacific Philosophical Quarterly*, 98, pp. 49–69.

Milam, P. 2019, "Reasons to Forgive," *Analysis*, 79(2), pp. 242–51.

Murphy, J. G. 2003, *Getting Even: Forgiveness and Its Limits*, New York: Oxford University Press.

Murphy, J. G., and J. Hampton 1988, *Forgiveness and Mercy*, Cambridge: Cambridge University Press.

Nelkin, D. 2013, "Freedom and Forgiveness," in I. Haji and J. Caouette, eds., *Free Will and Moral Responsibility*, Newcastle: Cambridge Scholars Publishing, pp. 165–88.

Norlock, K. J. 2009, "Why Self-Forgiveness Needs Third-Party Forgiveness," in S. Bloch Shulman and D. White, eds., *Forgiveness: Probing the Boundaries*, Oxford: Inter-Disciplinary Press.

Novitz, D. 1998, "Forgiveness and Self-respect," *Philosophy and Phenomenological Research*, 58, pp. 299–315.

Owens, D. 2012, *Shaping the Normative Landscape*, Oxford: Oxford University Press.

Pettigrove, G. 2004, "The Forgiveness We Speak: The Illocutionary Force of Forgiving," *Southern Journal of Philosophy*, XLII, pp. 371–92.

Pettigrove, G. 2009, "The Standing to Forgive," *The Monist*, 92(4), pp. 583–603.

Pettigrove, G. 2012, *Forgiveness and Love*, Oxford: Oxford University Press.

Radzik, L. 2009, *Making Amends: Atonement in Morality, Law, and Politics*, New York: Oxford University Press.

Richards, N. 1988, "Forgiveness," *Ethics*, 99:1, pp. 77–97.

Sabini, J., and Silver, M. 1982, *Moralities of everyday life*, Oxford: Oxford University Press.

Searle, J. R. 1979, *Expression and Meaning: Studies in the Theory of Speech Acts*, Cambridge: Cambridge University Press.

Searle, J. R., and Vanderveken, D. 1985, *Foundations of Illocutionary Logic*, Cambridge: Cambridge University Press.

Swinburne, R. 1989, *Responsibility and Atonement*, Oxford: Clarendon Press.

Walker, M. U. 2013, "Third Parties and the Social Scaffolding of Forgiveness," *Journal of Religious Ethics*, 41(3), pp. 495–512.

Warmke, B. 2015, "Articulate forgiveness and normative constraints," *Canadian Journal of Philosophy*, 45(4), pp. 490–514.

Warmke, B. 2016, "The Normative Significance of Forgiveness," *Australasian Journal of Philosophy*, 94(4), pp. 687–703.

Warmke, B., and M. McKenna 2013, "Moral Responsibility, Forgiveness, and Conversation," in I. Haji and J. Caouette, eds., *Free Will and Moral Responsibility*, Newcastle: Cambridge Scholars Publishing.

Zaibert, L. 2009, "The Paradox of Forgiveness," *Journal of Moral Philosophy*, 6, pp. 365–93.

Zaragoza, K. 2012, "Forgiveness and Standing," *Philosophy and Phenomenological Research*, LXXXIV(3), pp. 604–21.

10
Offsetting Harm

Michael Deigan

I increase the amount of CO_2 in the atmosphere by driving, buying plane tickets, ordering goods to be delivered to me, and so on. And the more CO_2 in the atmosphere, roughly, the worse various climate-change-related harms will be. So I worsen or, in other words, increase these harms.[1] Aside from polluting, we have many other opportunities to contribute to processes which result in more or less harm, depending on how we contribute. We can increase demand for factory-farmed animal products, for example, or add to social pile-ons. Given their ubiquity, it is important to know how we can permissibly interact with such processes.

Limiting our attention to cases where we increase some harm by a perceptible amount, we might think that there is no special theoretical difficulty here.[2] Increasing harm, after all, is just a way of doing harm. So we could reasonably hope to build an ethical theory around less complicated, more direct cases of doing harm, then apply this theory to the practically significant cases of merely increasing harm. But I think this is a mistake. There's a way of making some harm increases permissible that is not available in simpler cases of doing harm, one that we are bound to miss if we only consider the simpler cases.

Besides being able to contribute to harmful processes in ways that increase the harms that result, sometimes we can also contribute to them in ways that decrease those harms. We can buy carbon offsets, promote veganism or pay others to not buy animal products, reduce what others add to social harms, and so on. This means that sometimes we can act in ways that increase a harm, but still have total contributions *to that very harm* that

---

[1] See Broome (2019) for a recent argument against denials of this unfortunate fact.

[2] Much has been said about cases where one contributes to some collective harm in a way that seems to make no perceptible difference (Nefsky 2019). Though there may be overlap in applications, the harm increases I will be discussing are ones that I will assume do make a perceptible difference, and so do not face the same kind of inefficacy problems discussed in the collective/imperceptible harm literature.

Michael Deigan, *Offsetting Harm* In: *Oxford Studies in Normative Ethics, Volume 12*. Edited by: Mark Timmons, Oxford University Press. © Michael Deigan 2022. DOI: 10.1093/oso/9780192868886.003.0011

are net neutral or negative. Often a harm increase seems permissible if it is *offset* in this way with an equal or larger harm decrease.

This chapter is an exploration of offsetting harm increases from the perspective of more-or-less common-sense deontological theory.[3] I will argue that the standard deontological constraint against doing harm cannot accommodate permissible offsetting and so should be replaced by a constraint that can.

1. Zap Offsetting Cases

Let's start with some artificially simple cases.

Innocent C will soon get a mildly painful zap from evil B's zapping machine. How much the zap will hurt is determined by how much weight is on a scale attached to the machine: the more weight on the scale, the more painful the zap. One can remove a weight from the scale, but only if one adds a weight of one's own or pays a large fee. There are no other ways to interfere with the zapping. Currently there are 7 lb. worth of weights on the scale. The bystander A is aware of these facts.

Now consider A's actions in the following four variants of the case.

ZAP NON-INTERFERENCE: A neither adds any weight to the scale nor pays the large fee to remove a weight, leaving the total weight on the scale at 7 lb.

A's inaction here seems permissible. Sometimes one is obligated to aid a stranger facing some impending harm, but assuming that the fee is sufficiently high, the differences in zap painfulness sufficiently small, and that A doesn't have any special obligations to C, they would not wrong C by not taking on a significant cost to themselves in order to reduce the weight on the scale.

---

[3] More specifically, I'll be assuming there are moderate constraints that sometimes require agents to not do something even if that would lead to a more valuable outcome, as well as options both of agent-sacrificing and agent-promoting kinds, and a morally relevant distinction between doing and allowing. These assumptions are very controversial, but commonly enough held for a project which takes them as starting points to be of interest, even to those like myself who have doubts about them. Foerster (2019) discusses related phenomena from Kantian and a handful of consequentialist perspectives.

ZAP INCREASE: *A* has a 1 lb. weight that they don't feel like holding onto anymore, so they place it on the scale without removing any of the weights that were already there, leaving the total weight on the scale at 8 lb.

In ZAP INCREASE, *A*'s act makes the harm to *C* worse. Sometimes increases to harms are permissible, but this one is clearly impermissible.

ZAP OFFSETTING (NEGATIVE): *A* is carrying a 1 lb. weight, but would prefer to have a 2 lb. weight. *A* places the 1 lb. weight on the scale, and removes a 2 lb. weight, leaving the total weight on the scale at 6 lb.

Any harm increase that comes from the addition of a weight is offset by the removal of a heavier weight. In this case, *A* makes things better for *C* by making the trade. Given that they had no way to make things even better for *C* without taking on significant costs to themselves, this seems permissible.

Is net harm reduction necessary for permissibility? I think not. Neutrality seems to suffice.

ZAP OFFSETTING: *A* has a 1 lb. weight that is painted red, but would prefer to have a blue one. In this scenario, there are only 1 lb. weights on the scale. *A* places the red weight on the scale and removes a blue one, leaving the total weight on the scale at 7 lb.

C seems to have no more complaint against *A* here than they do in ZAP NON-INTERFERENCE. It would have been nice of *A* to pay the large fee to make things better for *C*, but this is not morally required.

So it seems that one can permissibly do something that makes a harm worse, such as adding a weight to the zapping machine scale, if one offsets it by reducing the harm by at least as much, such as by removing a different weight from the scale. The question I'm interested in is why increasing harm is permissible when paired with offsetting of this kind.

A straightforward answer seems to be available to those who go in for a familiar deontological constraint against doing harm. It's impermissible to do harm, except in special circumstances. In ZAP INCREASE, *A* does harm to *C* without good enough reason, so acts impermissibly. But in the offsetting cases, *A* doesn't do any harm or violate any other constraints, so acts permissibly, just like in ZAP NON-INTERFERENCE. Simple as that, right?

Wrong. This straightforward account fails, since one *does* do harm in offsetting cases (Section 2). Moreover, the permissibility of this harm cannot be captured by the standard provisos to the constraint against harm, such as the proviso that one may do harm if it is required for bringing about enough good (Section 3).

What I take the permissibility of offsetting cases to show is that the distinction that really matters is not one between harmful actions and non-harmful ones, but between actions which involve *unoffset harm increases* and those that don't. Doing something that increases a harm is not itself even pro tanto wrong; making a total contribution to it that is net positive is. I thus propose we replace a constraint against doing harm with one against making unoffset harm increases (Sections 4–6).

If this is right, it should be of interest to those concerned with finding the best way to formulate a deontological theory: a cornerstone of that approach needs revision. This revision may require us to rethink the source of deontological constraints (Section 7). It is also of practical interest. Given that emitting CO_2 will worsen the harms of climate change, is it permissible, as John Broome (2012) argues, to emit CO_2 so long as one reduces the amount of CO_2 in the atmosphere by at least as much?[4] And given that buying animal products can increase demand for them, and so increase the amount of suffering animals, would it be permissible to purchase such products so long as one, say, pays other people not to buy as much as they would have otherwise?[5] Whether such behavior is permissible will depend on what makes offsetting permissible, and whether these cases have the relevant features (Section 8).

2. Offsetting Not Prevention

Let us return to the natural proposal about the permissibility of offsetting that I say we should reject: offsetting cases are permissible because they don't really involve doing harm, since what offsetting does is prevent an action that would

---

[4] As Broome notes, this is not as simple as having trees planted, since once trees die they typically decompose and return the carbon they've stored back into the atmosphere. There are ways, though, to artificially extract and store carbon more or less permanently, even if they are currently rather expensive (Herzog 2018, ch. 6). Broome himself recommends "preventative offsetting", which doesn't take carbon from the atmosphere, but rather prevents gas that would have been emitted from being emitted (Broome 2012, 87). Like Barry and Cullity (2022), I am somewhat skeptical that this kind of prevention is sufficient for permissible offsetting, but I leave discussion of this issue for elsewhere.

[5] The question of meat offsetting has been raised by MacAskill (2015, ch. 8), who finds carbon offsetting permissible, but not adultery or meat offsetting.

have been harmful from being harmful. Broome, in his discussion of offsetting one's greenhouse gas emissions, suggests we explain its permissibility in this way. He says that offsetting is a way of doing "no harm" (Broome 2012, 85), so satisfies one's duty not to harm.[6] MacAskill also takes this approach. In contrast with attempting to offset one's adultery by buying indulgences, he says, "through effective carbon offsetting, you're preventing anyone being harmed by your emissions in the first place" (MacAskill 2015, 140).

This account of offsetting's permissibility relies on a couple of things: (i) an appeal to a standard constraint against doing harm, and (ii) the observation that there's no such constraint against doing things that would have been harmful were it not for some further action one takes to prevent that harm. Though (i) is controversial, it's widely enough held and defended that relying on it does not seem like a serious cost. In support of (ii), we can cite various cases where one can permissibly do something that would be harmful were it not for some other action one has performed or would go on to perform.[7] Suppose I pick up a loaded gun, aim it at someone, and pull the trigger. Normally, pulling the trigger would be an act that is wrong because harmful. But in this case, it turns out, I had carefully unloaded the gun after picking it up, so the trigger pulling wasn't harmful. My unloading the gun prevented the trigger pulling from being harmful.

The components of the prevention account, then, are reasonably well supported. It also seems to be a good intuitive fit for the cases. After all, the end result of an offsetting case doesn't just leave the victim with the same amount of welfare they would have had; it leaves them either better off or else *exactly* as they would have been, irrelevant Cambridge changes aside. So how could one have done harm? One *would have* done harm, of course, had one not offset, but this just goes to show that what offsetting does is prevent something from being harmful.

Attractive as the prevention account of offsetting initially seems, it can't be right. The problem is that it would require that the offsetting be linked in the right way to the particular would-be harmful acts that the offsetter performs. I have to unload *my* gun to prevent myself from doing harm. If I unload someone else's gun, that might prevent someone else from doing

---

[6] In his response to Cripps (2016), he makes this clear: "If you emit at one place, and also prevent an equal quantity of emissions at another place, *you do no harm* because you do not change the global concentration [of greenhouse gas]. This is how offsetting works" (Broome 2016, 159), emphasis added.

[7] Though certain of these cases may well cause trouble for maintaining (i); see Hanna (2015a, 2015b).

harm—a good thing, to be sure—but it doesn't prevent my trigger-pullings from harming. So the prevention account requires that offsetting actions interfere in some way with the effects of the would-be harmful actions they are supposed to offset. But in many permissible offsetting cases, including our original cases, there is no such link to be found.

In ZAP OFFSETTING, the weight A removes is a *different* weight from the one they placed on the scale, one that was there before A came on the scene. Removing it doesn't affect at all what the weight A put on the scale does. It doesn't affect it any more than it affects what the other remaining weights do. The weight A added is still on the scale at the time of the zap, still making the zap worse than it would have been had the weight been absent. If A wanted to prevent their earlier action from harming, they should have made sure to remove the very weight they placed on the scale. But it seems there's no requirement to do this—removing one is just as good a way of offsetting as removing any other of equal or greater weight. Similarly for other cases. As Broome (2012, 85) himself notes, carbon offsetting "does not remove the very molecules that you emit," and the molecules of CO_2 one has emitted "will wreak their damage" (p. 89), even when one has fully offset one's emissions. Offsetting actions don't need to affect in any way the consequences of the very actions they offset. So offsetting cannot be understood as preventing the harm that one's other actions would have done.

The prevention account has enough intuitive pull that it's worth dwelling on this point. Let's consider ZAP OFFSETTING in more detail. A places a 1 lb. weight, call it w, on the scale. If A doesn't go on to remove some weight or otherwise offset, this has clearly harmed C—it makes the pain C suffers worse, just like in ZAP INCREASE.[8] And nothing about what the w does is changed by A's offsetting action of removing some other weight from the

---

[8] This is clearly harming on a counterfactual comparative account, since C would have been better off had A not placed the weight there. It's also true if we accept any plausible non-counterfactual comparative theory of harming. On a very flatfooted theory of harming as causing a harm, at least with reasonable assumptions about causation and the identity conditions of events, adding the weight doesn't harm, since there is no harmful event it causes, only one it contributes to. But this is just a problem for such a flat-footed theory. A more sophisticated version of the theory, such as the one Harman (2009) ends up with, will count the addition as a case of harming, since it causes C to "be in a particular bad state rather than a better state" (Harman 2009, 150). What about accounts like Foot's (Foot 1967; 1984) or more recently Woollard's (Woollard 2008; 2015), which classify actions as harmings depending on their role in some harmful sequence? This will depend on the details. Foot's theory is silent on these cases. Adding a single weight neither originates nor sustains, in the relevant sense, a harmful sequence, but nor does it seem to be a mere enabling or forbearance from prevention. One might easily extend the account, though, to include a classification of increasing or

scale. There remains a rather direct causal path we can follow from the action to the worse pain. A puts w on the scale; w exerts downward force on the scale, which at the relevant moment contributes to higher voltage just like the other weights on it do, which makes for worse pain. Ordinarily, being able to trace such a causal path suffices for determining causation. And there's no preemption, overdetermination, or other factors which might undermine thinking w, like each of the other weights, is playing a causal role in the increasing the voltage: as with each of the other weights, if w were not there, the voltage would have been lower. The presence of w increases the harm to C, and it does so whether or not A has removed some other weight. So how could putting w on the scale not, in the end, harm C? I don't see a way of plausibly denying that it does. I conclude the prevention account fails to explain the permissibility of offsetting.[9]

3. Offsetting Not an Old Proviso

Even if I am right that offsetting cases involve doing harm, this does not yet show that permissible offsetting is incompatible with a constraint against doing harm. Practically nobody thinks that *every* possible harmful action is impermissible. If, somehow, the only way to prevent 1,000 innocent people from being killed is to punch some other innocent person, the punching is harmful yet permissible. It is also often permissible for surgeons to operate,

---

encouraging a harmful sequence as a case of harming. Woollard's account would treat the addition as a case of harming, since it is part of the sequence leading to the harm to C. But if I understand it correctly, it would also treat the removal of a weight as a case of harming for the same reason, which is unacceptable. The account could be modified, though, along the lines in which Harman modifies the simple causal account by treating only worsenings of harms as harmings.

[9] We can also make this point by observing that, in certain cases of offsetting, the harm increasing action and everything in the causal chain between it and the harm to the victim can be intrinsically identical to that of a case with a harm increase that is not offset. Suppose that A removed the weight before adding their own (they are allowed to remove a weight without adding one, but if they don't go on to add one they must pay the large fee). The removal will make the scale have 6 rather than 7 lb. Then the addition brings it from 6 to 7 lb. This still seems like a permissible offsetting case. But compare the weight addition here with that of another case, where the scale started with 6 lb. to begin with. Here adding a weight and doing nothing after would be a harm. But this act and everything between it and the zap to C could be an intrinsic duplicate of what happened in the offsetting case. And it is plausible that if for two possible actions everything in the causal chain from the action through some harmful event is intrinsically identical, then one of these actions increases that harm iff the other does (cf. Paul and Hall 2013, especially ch. 3, sect. 4.3). So, since there is harm in the non-offsetting case, there is harm in the offsetting case. So offsetting can't be prevention of would-be harm.

even if they must do some harm in the process, so long as they expect the procedure to do enough good for the patient on the whole.

There are at least two options available for holding onto a constraint against harm while allowing that these harms are permissible.[10] One is to say that the constraint is not one simply against doing harm, but is rather against something more complicated, like doing harm that is unnecessary to prevent some much worse outcome.[11] Another option is to say that the constraint against harm is violated in cases like these, but that the reason it provides against doing the action can be *outweighed* by considerations of general goodness or benefit to the victim.[12] In either case, if a harmful act is necessary for bringing about some amount of goodness, either a large total amount or a potentially smaller amount to the victim of the harm, the act can thereby be permissible. And there are various other proposed provisos to the constraint against harm, such as that doing harm can be permissible if consented to, if deserved, and so on. Why not think offsetting cases can be subsumed under some well-known proviso?

Offsetting has features which make the plausible candidates unsuitable. We've seen from the cases we've considered that increasing harm and offsetting can be permissible even when the benefit from doing so, as opposed to not interfering at all, is negligible. This goes for benefits to the agent, to the victim, or in terms of impersonal total goodness. In ZAP OFFSETTING, there is no net benefit for C, a minor benefit for A, and no benefit to anyone else. So provisos which say you may harm if doing so is required for accomplishing some significant amount of good (or preventing some sufficient amount of bad)—either for oneself, for the victim, or in general—won't work. Compare: a doctor may permissibly amputate your leg without your permission in order to save your life, but, assuming your legs are of roughly equal value to you, they cannot amputate your right leg in order to save your left without your permission, even if they would get some moderate benefit from doing so. It would be permissible to do this only if there were a rather large benefit to you or to others.

Note also that when a harm is to be justified by some benefit, it must be *required* for producing that benefit. Or rather, it must be required that there is some harm that is at least as bad. A doctor cannot permissibly amputate

[10] See, e.g., Kagan (1998a, sects. 3.2–3.3).
[11] Or some subtler variant of this idea, like the Doctrine of Initial Justification from Kamm (2007, ch. 5).
[12] Cf. the Tradeoff Idea from Thomson (1990, 123).

your leg simply for their own modest benefit even if they go on to save your life through some unrelated procedure. But in the offsetting cases we've considered the beneficial harm decrease does not require the harm increase, since A could have instead paid the fee to remove a weight.

The most promising provisos for making sense of offsetting, which concern a harm's relation to some benefit, won't save the constraint against harm. Other familiar provisos won't either. There is no consent from the victims in the cases we've considered. An appeal to hypothetical consent on its own doesn't help, since we'd need some further account of why offsetting would be consented to, whereas, for example, having one leg amputated to save the other would not be. Moreover, it seems to me that the permissibility of increasing harm and fully offsetting it doesn't change even with explicit non-consent from the victims of the harm, in which case a hypothetical consent proviso can't help. And as we saw in Section 2, there need not be any causal connection between the harm decrease and the harm increase it offsets, so we can't treat offsetting cases as ones of withdrawing one's own aid.[13]

Perhaps there are other provisos which have been proposed which could explain how the permissibility of offsetting is compatible with a constraint against doing harm. But none that I am aware of can, so I think we should explore other options. I hope, at any rate, that the force of the puzzle I am interested in can now be felt. Offsetting cases involve doing harm and so violate the familiar deontological constraint against doing harm, and not in a way covered by standard provisos. So why are they permissible?

4. A Recipe for Offsetting

It will help to first think about what makes an offsetting case an offsetting case. Not all the features of our two cases are essential. Instead of removing a 1 lb. weight, as in ZAP OFFSETTING, one could just as permissibly offset the weight addition by tying a 1 lb. balloon to the scale. And there's nothing special about weights and zaps, of course. A acts permissibly, for example, in the following.

---

[13] For discussion of withdrawing aid, see McMahan (1993) and references therein, as well as Woollard and Howard-Snyder (2021, sect. 7) for an overview of the more recent literature.

POISON OFFSETTING: C will get a headache from drinking out of a well which B has poisoned. How painful it will be depends on the amount of poison in the well. Currently there are 100 mg of it. A adds 10 mg of poison and an enzyme that will neutralize at least 10 mg of the poison.[14] So what features of offsetting cases are the essential ones?

Let's start with the obvious. One first needs some action that, at least on its own, would make some harm worse (adding a weight to the scale in ZAP OFFSETTING, adding poison to the well in POISON OFFSETTING). And one needs an action by the same agent that ensures that their total effect on the victim leaves them no worse off than they would have been had the agent not interfered at all (removing a weight, adding a balloon, adding the neutralizing enzyme).

As we can see from ZAP REPLACEMENT, however, the characterization thus far will not suffice.

ZAP REPLACEMENT: A prevents B from zapping C, but then goes on to zap C with the same voltage.

Here A's total behavior leaves C just as well-off as they would have been, but their zapping C seems impermissible. So we'll need a bit more complicated characterization to exclude this and similar cases from counting as offsetting.

I take the general structure of offsetting cases to be roughly this: independently of what some agent does, there's an impending intrinsically harmful event.[15] How bad it will be causally (not constitutively) depends on the magnitude of some quantity at some time: the more of the quantity at that time, the worse the harm. Before that time, the agent can alter that quantity both in ways that increase it and in ways that decrease it. Offsetting occurs when the agent does something to increase the magnitude of the

---

[14] Shouldn't A have just added the enzyme and not the poison? Suppose that A gets $1000 for adding poison, but must pay $950 for the enzyme.

[15] I'll be speaking of harms as events. This is controversial (Bennett 1995, ch. 2; Hanser 2008, 2011; and Thomson 2011), but not in ways that should matter here. An event is intrinsically harmful to someone if that event constitutively contributes negatively to that person's welfare. This is meant to be analogous to the more familiar notion of intrinsic value (or disvalue) in the 'as an end' sense, rather than the 'dependent only on intrinsic properties' sense (see Korsgaard 1983, Kagan 1998b, among others). I'll not theorize about 'impending' and 'independently' here, but instead leave them with their ordinary meanings. A more fully developed theory would require more precise characterizations.

quantity, but also does something that decreases it by at least as much, before the relevant time. The result is that the harm is not worse than it would have been had the agent not interfered.

This new characterization still applies to the offsetting cases. In ZAP OFFSETTING, the intrinsically harmful event is the pain C will have from B's zap, which was going to occur independently of what A does.[16] The quantity is the weight on the scale, which A can increase by adding weights and decrease by removing weights. In POISON OFFSETTING, the intrinsically harmful event is the pain from the headache C will get from drinking the poisoned water and the quantity is the amount of unneutralized poison in the well, which A can increase by adding more poison and decrease by adding the neutralizing enzyme. Having seen the pattern, it's easy to generate new offsetting cases: just plug in different kinds of impending harms, different kinds of determining quantities, and different ways for the agent to affect the quantity's magnitude.

In contrast with the offsetting cases, in ZAP REPLACEMENT there is no impending harmful event independent of what A does. The harm that comes to C—the pain from A's zap—comes entirely from what A does. So our recipe does not classify this as a case of offsetting, as desired.

Even if the recipe is the right one, though, it still doesn't yet tell us why increasing harm is permissible when offset. Let us now take up that question.

5. A New Constraint

Here's my suggestion: drop the familiar constraint against doing harm and replace it with a constraint against unoffset harm increases. If we do this, we can explain how the harms in ZAP OFFSETTING and other offsetting cases can be permissible, whereas the harm in ZAP INCREASE is not. The former will not violate this new constraint, whereas the latter does.

How would such a constraint look? One formulation just adds a conditional qualification to an ordinary moderate constraint against increasing harm:

CONDITIONAL OFFSETTING CONSTRAINT: Don't increase harm if you have not and will not fully offset it.

---

[16] To say that *this* event would have occurred regardless of what A does is to assume that events aren't especially modally fragile (on which, see Lewis 1986, 196–9).

But it will need to specify what it is to offset a particular harm increase.

Looking to our recipe for offsetting cases as a guide, we might propose the following. Where a is an act performed by an agent X that increases the badness of harm h solely through increasing an intermediary quantity q by degree d, a is fully offset iff there is some act β that X performs that decreases h through decreasing q by at least d. So in ZAP OFFSETTING, A's harm increase is fully offset because their removing another weight decreases the badness of the pain from the zap through removing as much weight from the scale as their harm increase added.

This proposal founders on slightly more complicated cases. Suppose in ZAP OFFSETTING A had added *two* 1 lb. weights but still only removes one 1 lb. weight. We want our constraint to rule out such behavior. A hasn't sufficiently offset here, and would need to remove 2 lb. worth of weights to do so. But on the current proposal, each of A's acts would be allowed. For each weight addition A makes, there is an act—namely, the single 1 lb. weight removal—that is sufficient for full offsetting according to the current proposal. The single harm decrease is being counted twice, offsetting two increases, each as large as the decrease is.

Trying to avoid this by linking specific increases to specific decreases runs into trouble when the increases and decreases vary in size. A better strategy is to take an individual's decreases to 'distribute' across all of their increases. So if they place a 1 lb. and a 2 lb. weight on the scale, a 1 lb. balloon offsets their first weight by ⅓ lb. and their second by ⅔ lb. To fully offset each harm, another 2 lb. worth of decrease is required.

Now that we are looking at total increases and decreases, though, an alternative formulation of the constraint seems simpler:

HOLISTIC OFFSETTING CONSTRAINT: Don't allow your net contribution to a harm be positive.

On this proposal, it's the whole pattern of actions that can violate the constraint, rather than any particular components of that pattern. It need not take on commitments about which increases are or are not fully offset. When A adds two 1 lb. weights but only removes 1 lb. worth, *Holistic Offsetting Constraint* doesn't say that either weight addition was a violation of the constraint. Instead, what violates the constraint is A's allowing their behavior as a whole to have the net effect that the harm is worse.

Though formulating the constraint this way takes us further from the familiar constraint against doing harm than *Conditional Offsetting*

Constraint did, there's nothing too exotic here. It's akin to what we might say about failing to keep a promise. What's wrong, we might say, is neither the act of promising itself nor any of the acts one performs instead of fulfilling the promise, but rather the pattern of behavior as a whole.

Holistic Offsetting Constraint gets the permissible offsetting cases right. In neither ZAP OFFSETTING nor POISON OFFSETTING does *A* make a positive net contribution to the harm. It also gets the partial offsetting cases right. However it's distributed across various actions, if *A* ends up adding more weight to the scale than they take away, *A* makes a positive net contribution to the harm, so runs afoul of the constraint. And it gets ZAP REPLACEMENT right. When *A* zaps *C*, there is a harm—the pain they caused *C*—to which they have made a positive net contribution, so they violate the constraint. This is so even if the net contribution of their behavior to *C*'s overall welfare is neutral or positive. What the constraint unoffset harm increases is sensitive to is not one's total contribution to a particular person's welfare, but rather one's total contribution to a particular harm.

Both the holistic formulation and the conditional formulation with proportional distributive linking seem to work for all the offsetting cases we've considered. There is, though, a potential problem remaining. A harm can be worse because it involves pain that is more intense, or because it lasts longer, or because it is more widespread; intensity, duration, and (apparent) spatial extent are all what I will call *aspects* of the harm that comes from pain. This means that a harm could be increased in one aspect and decreased in another. But just as one can't permissibly offset increases to one harm by decreasing another, as in ZAP REPLACEMENT, it seems to me that harm decreases can't offset harm increases to the different aspects of the same harm. If the pain of the zap were determined by two scales, one which determined the voltage of the zap and the other determined its length, I don't think it would be permissible for *A* to move weights from one to another without *C*'s consent, making the pain longer but less intense, even if the result is no worse than it would otherwise have been.

The current formulations of the constraint, however, don't make any distinction sensitive to this, and so would seem not to be violated by this kind of inter-aspect trade-off. Thus I suggest we revise the constraint to be either

CONDITIONAL OFFSETTING CONSTRAINT (ASPECTUAL):   Don't increase any aspect of harm if you have not and will not fully offset it, or

Holistic Offsetting Constraint (Aspectual): Don't allow your net contribution to any aspect of a harm be positive.

Though a full theory including such a constraint would require an account of what exactly an aspect of a harm is,[17] not to mention an account of what exactly a harm is, I think we can already see that it will sort the relevant cases correctly, and so is a promising way to explain the permissibility of offsetting.

6. Deriving What the Old Constraint Gets Right

The old constraint against harm is popular for a reason: there are plenty of ordinary cases of harming that it correctly rules out as impermissible. If we are going to replace the old constraint with a new one, we'd better make sure that we can still predict the wrongness of ordinary harms.

Happily, we don't need to make any modifications to the new constraint to do this, since it is violated by ordinary harmful actions. If I punch somebody, I make a net positive contribution to all of the aspects of the harm that comes to them from being punched. After all, I contribute all of the harm, and there's certainly a positive amount there. Could fully offsetting it make my harm increase in such a case permissible? Well, no. But this is no problem for the proposed derivation, since the reason offsetting can't make it permissible isn't because offsetting fails to do the normative work it does in the permissible offsetting cases, but rather because it's impossible in this case to fully offset the harm increase but still have done harm. If the increase to harm had been fully offset, there would have been no harm at all.

A harm of the familiar kind, then, is also a harm increase that is not fully offset. Thus—putting aside cases where offsetting happens—any violation of the old constraint against harm will also be a violation of the new constraint against unoffset harm increases, so we are not losing what the

[17] Here is a first pass: an aspect of a harm is a trope the existence of which is an ultimate normative partial ground of the fact that the harm is as intrinsically harmful as it is. But even granting this talk of ultimate normative grounding (see Fine 2012 and Bader 2017, but also Berker 2017), this only gets us close to what we're after. It makes a pain event's being a half-second long an aspect of the harm, whereas what we want is for the event's *temporal length* to be an aspect. I leave patching this up for elsewhere.

old constraint got right by moving to the new constraint. We should replace the traditional constraint against harm with a constraint against unoffset harm increases.

Sometimes the problem with a moral theory is that it misses a morally relevant distinction entirely, sometimes that it gives weight to a distinction not worth caring about. But other times the problem is more subtle: the theory has drawn a distinction which is *close* to a morally important one, but isn't quite carving at the moral joints. Theories that make this kind of error can be very plausible, since they may classify plenty of cases correctly, even if not for exactly the right reasons. But chances are they'll go wrong somewhere, potentially in very significant ways.

I've argued that theories which incorporate the usual kind of constraint against doing harm are making this kind of error. In correctly ruling out most cases of harm as impermissible, they overshoot, also ruling out cases of fully offset harm increases, like *A*'s action in ZAP OFFSETTING. A theory with a constraint against unoffset harm increases does better.

The remainder of the chapter briefly explores two kinds of implications of a shift to the kind of constraint I prefer.

7. The Source of the Constraint

Ultimately we want an account not just of what deontic constraints there are, but also an account of why there are the constraints that there are. Moving from a constraint against harm to one against unoffset harm increases may require us to rethink where this constraint is coming from.

An attractive idea to many deontologists is that autonomy is the source of a constraint against doing (as opposed to allowing) harm. For a person to have genuine free control over their own mind and body, they should be protected against certain kinds of impositions. On the one hand, they should be protected against being causally imposed upon by others in certain ways. This is why there is a constraint against doing harm.[18] On the other hand, a person also needs to be free from certain kinds of normative impositions, like being morally required to intervene on another's behalf. This is why there is no constraint against allowing harms.[19] I am cutting a long and

---

[18] See Quinn (1989), Shiffrin (2012), and Woollard (2015, ch. 6).
[19] For discussion, see Slote (1985, 23–34), Kagan (1989, 236–41), Shiffrin (1991), and Woollard (2015, 107–11).

controversial story short here, since my aim isn't to work through any detailed version of this view, but just to raise the question of whether something like it can be used to derive a constraint against unoffset harm increases.

It may seem easy enough for an autonomy theorist to explain why increasing harm and offsetting it would be permissible. A's behavior as a whole in ZAP OFFSETTING leaves C exactly as well off as they would have been had A not done anything to interfere. And it's not just that there's no overall difference in welfare for C, it's that there's no negative difference at all in the harm C suffers, but only irrelevant Cambridge changes or, in cases like ZAP OFFSETTING (NEGATIVE), improvements. And since A's not interfering at all would not impose upon C, it's plausible that C has not been imposed upon by A when A increases but fully offsets the harm, either.[20] And since there is no such imposition, it would be an unjustified normative imposition on A to require them not to behave in this way. So what A did in the offsetting case should be permissible.

But we are not done yet. We need to explain not only why doing harm in offsetting cases is permissible but also why doing harm is impermissible in others, including cases where the harm one does replaces another harm or lack of benefit of the same size, as in ZAP REPLACEMENT or cases of harming followed or preceded by compensation. This is more difficult, since it seems we can make a very similar argument for the permissibility of harm replacement. What A does in the replacement cases leaves C just as well off as they would have been had A done nothing. So how has C been imposed upon here any more than in the offsetting cases?

One could say that while all of A's actions together don't make C worse off, one of them (the zap) does, which is enough of an imposition to justify making it impermissible. But this won't do, since the same point could be made about the offsetting case if we single out the harm increasing act.

We might try appealing to the point that, in offsetting cases, it's not just "no difference in the victim's welfare" but "no difference at all in the harm C suffers, improvements and irrelevant Cambridge changes aside". Even if you should be indifferent between not being zapped and being zapped but given $200, it's not implausible to say that it's a violation of your autonomy if I exchange one for the other without your permission. But what about ZAP REPLACEMENT? We can make the harms in question intrinsically identical,

---

[20] Note that we need to be considering the agent's behavior as a whole here, rather than its proper parts (cf. Portmore 2017 and Brown 2018).

not just equal in badness, so how can replacing one with the other be a morally relevant imposition?

It seems we need some way of explaining why the identities of harms would be important: why replacing this harm with that qualitatively identical one would be impermissible, whereas adding and subtracting the same amount from it can be permissible. It is unclear how an appeal autonomy can make sense of this.

This is not to say that we should give up on autonomy as a source of the constraint against unoffset harm increases. I think it remains a promising direction to pursue. But we may find that some alternative theory does better. In any case, more work is required to derive the constraint against unoffset harm increases from deeper principles, and current frameworks may require revision or replacement in order to do so. Moving to such a constraint may thus have theoretical implications that go beyond questions about how best to formulate a first-order deontic theory.

8. CO_2 Emissions and Meat Purchases

Moving to a constraint against unoffset harm increases may also make a difference to what is permissible in practically important, real-life cases.

In the climate change case, it's plausible that the harms from emitting CO_2 are ones which come solely from increase to the global concentration in CO_2, as Broome (2012, 2016) claims. And the global concentration of CO_2 is a quantity that one can increase or decrease. One could worsen the harms of climate change by emitting CO_2, but also reduce those very harms by offsetting those emissions. So it appears to fit our recipe for offsetting cases, opening up the possibility of permissible offsetting, though there will be tricky issues about timing and the identity of harms we would need to consider. The question will be whether by emitting and buying carbon offsets one's behavior will involve a net increase to any aspect of any harm.

In the animal product case, it is plausible that the most significant harms are the ones that come solely from increased demand for these products from specific suppliers within specific windows of time. Plausibly, it is only the amount of demand for these products as perceived by the relevant suppliers at the relevant times which determines how bad the harms are. One can increase this demand through purchasing, but perhaps there are things one could do—paying others not to purchase, for example—to decrease this demand.[21] So it seems plausible that this might fit the offsetting recipe, again opening up the possibility of permissible offsetting.

It would take a good deal of empirical and philosophical work to sort out the details of these cases. We cannot hope to settle here whether they could involve permissible offsetting and, if so, how exactly that offsetting can be done. We have laid some important groundwork for approaching these questions, however. If what I've argued is right, what we need to look for is not whether any potential means of offsetting will prevent our actions from harming or help them meet any of the usual provisos to a constraint against harm. Rather, we must look to see whether our behavior as a whole will have the net effect of worsening a harm in any way. If it does, we act (pro tanto) wrongly. It may be, though, that we can act in ways that worsen harms but act in other ways which offset those worsenings. In this case our response, other duties aside, need not be to refrain from these harmful actions, but rather to ensure they are fully offset.[22]

References

Bader, Ralf. 2017. "The Grounding Argument Against Non-Reductive Moral Realism." In *Oxford Studies in Metaethics*, ed. Russ Shafer-Landau. Vol. 12. Oxford University Press.

Barry, Christian and Garrett Cullity. "Offsetting and Risk Imposition." *Ethics* 132 (2): 352–81.

Bennett, Jonathan. 1995. *The Act Itself.* Clarendon Press.

Berker, Selim. 2017. "The Unity of Grounding." *Mind* 127: 729–77.

Broome, John. 2012. *Climate Matters: Ethics in a Warming World.* W. W. Norton & Co.

---

[21] This would be difficult to implement. One would need to know that someone was going to make a particular kind of purchase, that paying them would keep them from making that purchase, that they will not just buy something else objectionable instead, and that they weren't disposed to make the purchase in the first place because they expected you to pay them not to. Similar complications go for preventative offsetting of carbon emissions.

[22] One may have duties from other sources to do more than this. Besides offsetting one's own emissions, one may be morally required to push for international political solutions. Besides offsetting animal product consumption, one may be required to promote laws requiring humane treatment or advocate that others offset their consumption as well. And there may be additional duties to not participate at all, such as a duty to avoid complicity through benefiting from wrongdoing (McPherson 2018).

Broome, John. 2016. "A Reply to My Critics." *Midwest Studies in Philosophy* 40: 158–71.

Broome, John. 2019. "Against Denialism." *The Monist* 102: 110–29.

Brown, Campbell. 2018. "Maximalism and the Structure of Acts." *Noûs* 52 (4): 752–71.

Cripps, Elizabeth. 2016. "On *Climate Matters*: Offsetting, Population, and Justice." *Midwest Studies in Philosophy* 40: 114–28.

Fine, Kit. 2012. "Guide to Ground." In *Metaphysical Grounding: Understanding the Structure of Reality*, ed. Fabrice Correia and Benjamin Schnieder, 37–80. Cambridge University Press.

Foerster, Thomas. 2019. "Moral Offsetting." *The Philosophical Quarterly* 69 (276): 617–35.

Foot, Philippa. 1967. "The Problem of Abortion and the Doctrine of Double Effect." *Oxford Review* 5: 5–15.

Foot, Philippa. 1984. "Killing and Letting Die." In *Abortion and Legal Perspectives*, ed. Jay L. Garfield and Patricia Hennessey. University of Massachusetts Press.

Hanna, Jason. 2015a. "Doing, Allowing, and the Moral Relevance of the Past." *Journal of Moral Philosophy* 12: 677–98.

Hanna, Jason. 2015b. "Enabling Harm, Doing Harm, and Undoing One's Own Behavior." *Ethics* 126: 68–90.

Hanser, Matthew. 2008. "The Metaphysics of Harm." *Philosophy and Phenomenological Research* 77: 421–50.

Hanser, Matthew. 2011. "Still More on the Metaphysics of Harm." *Philosophy and Phenomenological Research* 82: 459–69.

Harman, Elizabeth. 2009. "Harming as Causing Harm." In *Harming Future Persons*, ed. Melinda A. Roberts and David T. Wasserman, 137–54. Springer.

Herzog, Howard J. 2018. *Carbon Capture*. MIT Press.

Kagan, Shelly. 1989. *The Limits of Morality*. Clarendon Press.

Kagan, Shelly. 1998a. *Normative Ethics*. Westview Press.

Kagan, Shelly. 1998b. "Rethinking Intrinsic Value." *Journal of Ethics* 2 (4): 277–97.

Kamm, F. M. 2007. *Intricate Ethics: Rights, Responsibilities, and Permissible Harms*. Oxford University Press.

Korsgaard, Christine M. 1983. "Two Distinctions in Goodness." *Philosophical Review* 92 (2): 169–95.

Lewis, David. 1986. "Events." In *Philosophical Papers*, Vol. II, ed. David Lewis, 241–69. Oxford University Press.

MacAskill, William. 2015. *Doing Good Better*. Penguin Random House.

McMahan, Jeff. 1993. "Killing, Letting Die, and Withdrawing Aid." *Ethics* 103: 250–79.

McPherson, Tristram. 2018. "The Ethical Basis for Veganism." In *The Oxford Handbook of Food Ethics*, ed. Anne Barnhill, Mark Budolfson, and Tyler Doggett, 210–40. Oxford University Press.

Nefsky, Julia. 2019. "Collective Harm and the Inefficacy Problem." *Philosophy Compass* 14 (4).

Paul, L. A., and Ned Hall. 2013. *Causation: A User's Guide*. Oxford University Press.

Portmore, Douglas W. 2017. "Maximalism Versus Omnism About Permissibility." *Pacific Philosophical Quarterly* 98 (S1): 427–52.

Quinn, Warren S. 1989. "Actions, Intentions, and Consequences: The Doctrine of Doing and Allowing." *Philosophical Review* 98 (3): 287–312.

Shiffrin, Seana Valentine. 1991. "Moral Autonomy and Agent-Centered Options." *Analysis* 51 (4): 244–54.

Shiffrin, Seana Valentine. 2012. "Harm and Its Moral Significance." *Legal Theory*, 18(3): 357–98.

Slote, Michael. 1985. *Common-Sense Morality and Consequentialism*. Routledge & Kegan Paul.

Thomson, Judith Jarvis. 1990. *The Realm of Rights*. Harvard University Press.

Thomson, Judith Jarvis. 2011. "More on the Metaphysics of Harm." *Philosophy and Phenomenological Research* 82: 436–58.

Woollard, Fiona. 2008. "Doing and Allowing, Threats and Sequences." *Pacific Philosophical Quarterly* 89: 261–77.

Woollard, Fiona. 2015. *Doing and Allowing Harm*. Oxford University Press.

Woollard, Fiona, and Frances Howard-Snyder. 2021. "Doing Vs. Allowing Harm." In *The Stanford Encyclopedia of Philosophy*, ed. Edward N. Zalta, Fall 2021. https://plato.stanford.edu/archives/fall2021/entries/doing-allowing/.

11

Please Keep Your Charity out of My Agency

Paternalism and the Participant Stance

Sarah McGrath

1. Introduction

It is a familiar fact that when it comes to interpreting what someone is saying, you have some choices: different circumstances call for different strategies. Sometimes it is best to:

(i) take what they said at face value, as something to be engaged with on its own terms,

and other times it is best to:

(ii) write it off as the product of non-rationalizing causes.

There are cases in which it is obvious that you should opt for (i) over (ii). When your child tells you "I'm hungry" around lunchtime, you take this at face value, and make him some lunch. There are cases in which it is obvious that you should go for (ii) over (i). When he screams "I AM NEVER BIKE RIDING AGAIN!" after skinning his knees in a bike crash, you think that, probably, it's the skinned knees talking. Rather than putting his bike up for sale on eBay, you kiss and bandage the knees. And, of course, there are many cases where the correct interpretation is somewhere in between.

The choice we have about different ways of interpreting people that I have just presented is related to distinctions drawn in philosophical discussions of determinism and moral responsibility. Strawson (1962) draws a distinction

Sarah McGrath, *Please Keep Your Charity out of My Agency: Paternalism and the Participant Stance* In: *Oxford Studies in Normative Ethics, Volume 12*. Edited by: Mark Timmons, Oxford University Press. © Sarah McGrath 2022.
DOI: 10.1093/oso/9780192868886.003.0012

between the "participant" and "objective" attitudes that we can take toward others: to take the participant attitude is to react to "the good or ill will or indifference of others towards us, as displayed in their attitudes and actions." But when we see someone's "deranged or compulsive" behavior as a product of "peculiarly unfortunate . . . formative circumstances," we step out of the participant stance, and (at any rate if we are among "the civilized") take an objective attitude toward him. Importantly, taking an objective attitude and taking the participant stance are not mutually exclusive: even in the same situation, Strawson points out, we can take both the objective attitude and reactive attitudes, as these are "not altogether *exclusive* of each other . . . " (Strawson 1962: 194).[1]

Whereas Strawson is interested in the fact that we *can* take these different attitudes toward persons because he is interested in whether we can make sense of moral responsibility, given that persons are at the same time "things," I am interested in the normative ethical question of what we owe to other people in terms of how we interpret them. As Rae Langton points out in her discussion of Kant's correspondence with Maria von Herbert, it is very often *bad* to treat a person "as a thing" by regarding her behavior as the product of forces that cause but do not rationalize it (Langton 1992). Early in the correspondence, Kant regards von Herbert as a dear friend whose questions are worthy of serious consideration, but he later dismisses her as an "ecstatical little lady" who suffers from a "curious mental derangement" (Langton 1992: 500). Langton comments that "it is hard to imagine a more dramatic shift from the interactive stance to the objective," and describes Kant, in having made this shift, as having forgotten "an important aspect of the duty of respect, which requires something like a Davidsonian principle of charity" (Langton 1992: 500).

But plausibly it is not *always* disrespectful to think of a person as a "mere thing," and to take her behavior to be a product of non-rationalizing (or "mere") causes. Strawson says we suspend the participant attitudes when we encounter an agent who is "warped or deranged, neurotic or just a child"; we

---

[1] Nagel appeals to a similar contrast to explain why he thinks that there may be no solution to the problem of "Moral Luck" (1979). We can oscillate between an "external view" and an "internal view": to see someone as exercising agency and as a locus of moral responsibility is to take the internal view; to see her actions determined by a natural sequence of cause and effect is to take the external view. According to Nagel, these two "views" are fundamentally opposed: "something in the idea of agency is incompatible with actions being events, or people being things" (1979: 37).

also do so when we judge the agent "wasn't himself" or "has been under very great strain recently" or was "acting under post-hypnotic suggestion." We are all familiar with cases in which loved ones act badly out of anger or duress. In these cases, perhaps regarding a person's behavior as the product of something other than her agency, and not holding her responsible for it, is a kindness rather than an insult.

This normative question about interpersonal interpretation is discussed both in Langton (1992) and, more recently, in Mark Schroeder's "Persons as Things" (2019). Langton suggests in passing that respecting others can involve interpreting them charitably. On the view that Schroeder (somewhat tentatively) defends, we correctly deploy the causal interpretation of a person's behavior when doing so is charitable: when it "makes her contribution to the world greater, better, or more significant" than it would be, were her behavior attributed to her agency (2019: 109).

I think that this view has a lot going for it. But I worry that it amounts to a kind of interpretive analogue of paternalism. According to Shiffrin (2000), the problem with paternalist action is that it involves my substituting my judgment for yours, in a sphere of your legitimate agency, on the basis of my view that my judgment is *better* than yours. Paternalist actions thus "directly express insufficient respect for the underlying capacities of the autonomous agent" (Shiffrin 2000: 220). A policy of deploying the most charitable interpretation of the behavior of others might be objectionable in a similar way: if my interpretive norm instructs me to interpret you as making the *best* contribution that you could be making, I run the risk of showing you insufficient respect, assuming that your contribution is in some sense within the sphere of your legitimate agency.

In this chapter, I argue that the same considerations that tell against my acting paternalistically toward you tell in favor of a strong default policy of taking what you say at face value, as an undistorted expression of *you*. I argue that it is permissible to depart from this default only if there is some positive reason to think that distorting factors are at work. I take this answer to mirror the view in literature on the epistemology of testimony according to which the correct epistemological norm instructs us that we can trust the testimony of others, absent positive evidence that they are not to be trusted. On my view, we should take this default stance, even if the resulting interpretation is not the most charitable.

2. Persons, Things, and the Principle of Charity

At the beginning of "Persons as Things," Schroeder offers the following gloss on what it is to be "treated as a thing":

> To be treated as a thing is to be minimized, rather than engaged with, predicted and controlled rather than reasoned with, written off as the product of our environment rather than appreciated for our unique contributions.   (2019: 95)

The central example of objectionably treating a person as a thing that Schroeder discusses in the paper is this: his wife Maria comes home from work and compliments him on the work he has done cultivating some jasmine vines in their yard, but instead of being grateful for the compliment, he wonders whether she said that because she found a coin in a vending machine (2019: 97). To think of Maria's compliment as the effect of her finding a coin in a vending machine is to treat her as a thing, in a way that renders the compliment ineligible for gratitude: "it only makes sense to be grateful to someone within the participant stance – as part of a collection of attitudes that together constitute taking them to be a someone, and not just a something" (2019: 98). To understand her compliment as subject to "non-rational prediction and explanation by background facts" is to "diminish" her (2019: 97).

However, interpreting behavior as "subject to non-rational prediction and explanation" doesn't *always* diminish anyone. The central example that Schroeder offers to illustrate this point is:

> ... Maria comes home and snaps at me that my Jasmine vines are coming in too slowly ... Maria's behavior, ultimately, [is] better explained by an underlying bad mood than as a conversational move that requires a response.   (2019: 6)

Schroeder says that to see Maria's behavior as explained by a bad day at work rather than something to be engaged with on its own terms, and to simply to dismiss the negative comment as something that is not to be resented or responded to is *not* diminishing (2019: 101). More generally,

> deploying causal interpretation as an interpretive tool is actually one of the key elements of successful interpersonal engagement. In order to most successfully engage with another specifically as a person – or put

differently, in order to fully occupy the participant stance toward them – you must have the tool of causal interpretation at your disposal.

(2019: 105)

If in some cases causal interpretations are diminishing and in other cases they are not, then what explains the difference? When does interpreting someone's behavior as "merely caused"—thinking of them as a thing, with respect to that particular behavior—enable taking the participant stance toward that person, rather than diminishing her? Schroeder's answer is that interpreting someone's behavior as merely caused enables taking the participant stance whenever the causal interpretation is *more charitable*. If thinking of a particular piece of behavior as merely caused makes the person's overall contribution to the world *better*—as it does in the case of Maria's insult regarding the vines—then thinking of it in this way does not diminish her, and it is not morally objectionable. If it makes the overall contribution to the world seem worse—as in the case of the compliment—then it does diminish her, and it is morally objectionable (2019: 109).

Now, we might worry that this idea of the "best" interpretation of the person's "overall contribution to the world" is underspecified: best in what way exactly? Different interpretations might be better along different dimensions. Moreover, one might think that when it comes to interpreting texts, the most charitable interpretation of a given element cannot be determined independently of its place in the narrative or argumentative structure of the whole.[2] If something similar holds for the case of persons, then charitably interpreting a piece of behavior would also be a holistic matter—one that couldn't be carried out until the story has an ending and the life is complete. If only a fraction of the story has been written, then it is arguably indeterminate whether attributing a given action to the agent's true self is charitable, or not.[3]

But supposing that there is some satisfactory way of specifying an interpretive background and an account of what would make a contribution best, the next question is: what explains why the principle of charity would be the appropriate principle for guiding the relevant interpretive practice? Schroeder argues that this is the appropriate principle because "charity

---

[2] This certainly seems true for Schroeder's case of interpreting philosophical writing, discussed below.

[3] I am grateful to Renee Bolinger for this point.

tracks truth" (2019:108).[4] In other words, if you use the principle of charity, then you are more likely to interpret someone's behavior *accurately*. This is a surprising claim! Offhand, this might seem Panglossian: aren't a lot of us in fact less good than we could charitably be interpreted as being? Why think that most people are as good as they could be interpreted as being, as opposed to being somewhere in between the worst and the best?

Schroeder's answer appeals to an analogy with the case of philosophical writing. Presumably, we all agree that one ought to interpret the authors of philosophical writing charitably. The reason, Schroeder suggests, is that in doing so one is more likely to arrive at the correct interpretation of the piece of writing. The most charitable contribution is most likely to be correct because "the real contribution made by a particular piece of argumentative writing simply is the best, most charitable interpretation of that piece of writing – the one that allows its contribution to be greatest, given the background constraints" (2019: 108). This explanation of why we should read philosophical works charitably can be carried over to the case of persons. Just as the real contribution of a piece of philosophical writing is constituted by the most charitable reading, persons are "constituted by the best interpretation of what contribution their behavior makes to the world" (2019: 110). Since interpreting Maria's negative comment about the vines through a causal lens enables a more charitable interpretation of her behavior—one "that makes her contribution to the world greater, better, or more significant" (2019: 14)—the causal interpretation is correct.

The person-text is the totality of the person's behavior, but not all of her behavior plays the same role. For example, sleepwalking is "not properly attributable to the person in the fullest sense" (2019: 109). What about flat-out assertions? Schroeder says,

self-avowals must be treated with deference, but even self-avowals can go wrong…Whenever your self-avowals are best explained in causal terms, even those self-avowals are not properly yours, and so they do not place a hard constraint on which behavior belongs as part of the story about what contribution you are actually making.   (2019:109)

---

[4] I am grateful to an anonymous referee for pointing out that "charity tracks truth" can be read in a stronger and a weaker way. On the strong reading, most charitable interpretation is the true one. On the weaker reading, being charitable is a good-making feature of an interpretation, which tends to make it true but does not *entail* its truth. (Other bad-making features might outweigh it, making the interpretation untrue.) Schroeder's text seems to make clear that he intends the strong reading.

One nice feature of this account is that it can explain why giving a causal explanation of a compliment will typically require more evidence than giving a causal explanation of a complaint: "the range of circumstances in which a compliment helps to constitute a person's contribution to a relationship or shared project or to the world more generally is wider than that in which a complaint does so" (2019: 14). It can also explain why causal interpretations feel diminishing when they do.

There are two questions we can ask about this: (1) Are persons constituted by the most charitable interpretation of the totality of their behavior? (2) If persons are constituted by the most charitable interpretation of the totality of their behavior, should we use charity as our interpretive norm? In the next section, I argue that even if it turned out that the answer to question (1) is "yes", the answer to (2) is still "no", because interpreting someone charitably can be wrong for the same kinds of reasons that paternalism can be wrong.

3. Paternalism, Charity and the Constitution Claim

Could it ever be paternalistic to interpret someone charitably? Consider the following example (from Shiffrin 2000):

> ... an interlocutor raises his hand at a talk. He is called upon and just as he haltingly begins to articulate his point, an excited, sympathetic colleague loses self-control and interjects: "Isn't this a better way to put the point?" She goes on to drown him out while cleverly and eloquently articulating his point. She takes over his question because she feels she has a better command of it than he does.   (2000: 217)

Let us agree with Shiffrin that the sympathetic colleague acts paternalistically. Shiffrin uses this example to make the point that behavior can be paternalistic even if it is not aiming at anybody's welfare: the sympathetic colleague is trying to get the best version of the question into the discussion, rather than trying to benefit the speaker. But I think that the example can also be used to motivate the claim that charitably interpreting someone can be paternalistic: we can see the interrupting colleague as being charitable in that she attributes the *best* interpretation of the question to the speaker.

It might be objected that even if it is true that the senior colleague *acts* paternalistically, this doesn't show that *interpreting* someone charitably can be paternalistic: what was paternalistic is not the interpretation itself but, rather, interrupting and speaking out of turn. When one person legitimately has the floor, and another takes over with a "helpful" outburst, *that* is what is diminishing and disrespectful.

But I think that if a questioner has authority over what her question is, it can be paternalistic to interpret her charitably, even if this does not involve speaking at all. The sympathetic colleague could just think, "yeah that is not what she meant to ask." And that feels like a paternalistic thought, even if the senior colleague does not say anything at all.

As a warm-up to the idea that an interpretation can be paternalistic, let's start with the fact that it is clearly possible to be paternalistic by *omission*. Example: you get a text that says that the talk has been delayed. But you don't tell me, because if I don't know, then my chronic inability to correctly estimate how long it will take for me to get ready won't derail us, and we will actually be on time. Paternalism-by-omission cases show that being paternalistic does not require any action on the part of the paternalist. If what unifies the action and omission cases is their motive, it seems plausible that an interpretation with the right kind of motive could also count as paternalistic, or at least count as bad for the reasons that paternalism is bad.[5]

It might be tempting to think that if charitable interpretations can be wrongfully paternalistic, then we should reject:

Constitution Claim: The most charitable interpretation in light of a person's behavior is constitutive of the person.

The argument might look something like this:

1. Charitable interpretations can be paternalistic.
2. Paternalistic interpretations do not aim at the truth about what a person thinks, means, and says.

---

[5] To be clear, I am not saying that it is *always* paternalistic to interpret someone charitably, nor that interpreting someone paternalistically is always wrong. Maybe interpreting my 6-year-old charitably is paternalistic but not wrong. Moreover, I do not mean to suggest that questioning someone's self-understanding from an external point of view is always problematic: your assessment of my reasons and motivations will often be a healthy corrective of my own assessment of those reasons and motivations, when our assessments come apart. I am grateful to A. K. Flowerree for this point.

Preliminary conclusion: Interpreting someone charitably can come apart from aiming at the truth about what she thinks, means, and says. Conclusion: It is not true that the charitable interpretation is constitutive of the person.

In other words: if the Constitution Claim is true, then by using the principle of charity in interpreting someone, you would simply be trying to find out the true interpretation, because *charity tracks truth*. But if paternalistic interpretations don't aim at the truth, then there is pressure to reject the Constitution Claim.

But this argument is too quick. The friend of the Constitution Claim can reject premise 2. Recall the case of the bike crash that caused my 6-year-old, Hugh, to scream "I am never bike riding again!" At the moment when he says this, he is dead serious about the claim that he is quitting for good. Nevertheless, I do not take what he said at face value. Why not? Well, although there is a sense in which he "really means" it, there is also a sense in which he *doesn't*. Even if he means it in the heat of the moment, he won't hold this view tomorrow. Tomorrow he will be back in the saddle, trying new tricks. Taking a leaf from Rawls, we might say: *I will never ride a bike again* is not among his "considered" views.[6] So paternalistic interpretations can aim at *a* truth about a person: a truth about her best or considered self. Premise 2 is false when we read it as about that self. It is only plausible when read as a claim about what the speaker means in the heat of the moment. When I interpret you paternalistically, I am dismissive of what the less-than-ideal version of you is saying right now. The friend of Constitution Claim will not be worried by this, provided that her claim is about the considered or true self.

But once we have distinguished between two versions of a person—the "as-is" version (which might have had a rough day at work and be grouchy, mad, scared, drunk, scraped-kneed, frustrated, jealous, confused, etc.) and the considered or best version (the version that is not "off" in any of these ways and is therefore expressing her views that are stable over time)—I think that it becomes clear that charity cannot help us with the question with which we began. Perhaps it is true that charity helps you to figure out what long-term desires, intentions, and beliefs to attribute to the real "person text" or the true self. Returning to Schroeder's central example: when Maria

---

[6] Rawls deploys the term "considered judgment" in several places; for a representative characterization see Rawls (1974).

comes home from work and insults his jasmine vines, perhaps charity helps him to determine that this is not Maria's considered self. Her considered self would not insult the vines in this way! Charity tells him how to get from grumpy-Maria to Maria-considered. But neither the principle of charity nor the Constitution Claim tells Mark how to engage with what the grumpy-Maria before him just said. These principles are silent about whether what *she* said should be dismissed or engaged with on its own terms. Perhaps charity tells him how Maria-considered would behave, but this just leaves open whether what grumpy-Maria said can be ignored.

To summarize: if the Constitution Claim is about the as-is self, then the argument against it goes through, because charitable interpretations need not aim at what as-is people think, mean, or are saying. But if the Constitution Claim is about the considered or best self, it does not answer our original question. The promise of the principle of charity was to tell us when it is perfectly okay to write off someone's behavior as merely caused and opt out of engaging with it on its own terms. But *charity* doesn't tell you what to do about the behavior of the out-of-sorts, as-is version of the person before you. I myself think that writing off what the as-is version of a person is saying can be objectionably paternalistic. And paternalism, if not exactly the problem that we were worried about when we began, is clearly in the same ballpark.

4. Toward a Better View

The examples that we have been focusing on raise the question: when should you dismiss what your interlocutor is saying as the product of mere causes, and why? Since the issue concerns interpreting what somebody says to you, the literature on testimony might be instructive. A central question in this literature is that of why we are justified in relying on the testimony of others to the extent that we are. The majority view is that you don't need a positive reason to think that a speaker is reliable in order to take her at her word; rather, trust is the *default*. Unless you have a positive reason to doubt that the speaker is reliable, you are justified in believing her. You needn't have special reason to think that she is *not* confused or *not* a liar; you just need to *lack* evidence that she *is*. Thus, Burge writes:

> [A] person is entitled to accept as true something that is presented as true and that is intelligible to him, unless there are stronger reasons not to do so.
> (Burge 1993: 457)

One reason to think that this must be correct is that if one needed a positive reason to believe that one's sources were reliable, then we wouldn't have anywhere near as much knowledge as we think we have. The testimony of others is a source of a vast number of our beliefs about the past, the future, and the world around us, but in most cases we do nothing to verify that the information is correct or that the speaker can be relied upon. If beliefs acquired from testimony only counted as knowledge in cases where the hearer was in possession of evidence that a speaker could be trusted, then we would have to be skeptical about many domains in which we take ourselves to know a great deal.[7]

Now the "entitled" in Burge's principle is the "entitled" of epistemic entitlement. And our question is not about what a listener is epistemically entitled to do but, rather, about what we owe a speaker as matter of respect. Moreover, Burge's principle is about when we are entitled to take what the speaker said as *true*, whereas we are interested in when we should engage with it on its own terms, rather than writing it off. So we can't take this idea from the testimony literature wholesale. But we can borrow from it the idea of a default to trust the speaker, where the relevant kind of trust is trust that what she is saying should not be dismissed as the product of mere causes.[8] Let's consider a principle that makes this the default:

Default to Trust: Absent good reason to seek a merely causal explanation, don't.

Default to Trust issues the right result for a number of cases. For example, it says that Mark should engage with Maria's compliment about his vines on its own terms, not because doing so is more charitable but because there is no reason to seek a causal explanation. By contrast, I do not take Hugh's anti-biking vow at face value since I witnessed his terrible crash. And, depending on the details of the case, it might be that Mark should not "write off" the complaint about his vines: whether he should depends on the quality of his reason to think that the bad day at work undermines it.

We noted above that an advertised attraction of the charity view is that, if writing off a compliment typically requires more evidence than writing off a

---

[7] See Coady (1992). For similar arguments see also Foley (1994), Plantinga (1993), Strawson (1994), and Webb (1993).

[8] For the development of a similar default against dismissing the avowed reasons of one's interlocutor see Flowerree (Forthcoming).

complaint, charity can explain this. The explanation is that, as a general rule, seeing compliments as sincere tends to contribute positively to the interpretation of a person's behavior more often than seeing complaints as sincere does. But if Default to Trust is true, then writing off a compliment will not, in fact, require more evidence than writing of a complaint. But perhaps the friend of Default to Trust could just deny that this asymmetry is real, as opposed to a kind of wishful thinking or bias. True, we *prefer* compliments to complaints. Consider the typical professor encountering her teaching evaluations. Some of the comments students write are compliments, and some are complaints. If the complaints are easier to write off than the compliments, that might just be because, all things equal, complaints feel bad and compliments feel good.

One might wonder whether the charitable norm that tells Mark to write off the utterances of Maria as she is now when they are not what the considered version of Maria would have said is a norm that actually *conflicts* with Default to Trust. After all, as its name suggests, Default to Trust is a mere default principle. Couldn't one both give the most charitable interpretation and *also* have a default policy of taking her interlocutors at face value?

I believe that the two norms actually do conflict. First, the constitutive view is not just that you have *permission* to write off the comment about the vines, but that you would be making a mistake if you didn't. Since the considered version of Maria would not have said it, you should dismiss it. So Default to Trust does conflict with the norm that says to write off utterances of the as-is versions of a person whenever the best version of the person wouldn't have said it. If you are following the charitable interpretive norm, then, each time someone says something, you should be disposed to reach for the interpretation of this behavior that makes her true self out to be as good as it could be. And if you are doing that, then you are not giving special default status to the idea that she is managing to exercise her agency unadulterated: so long as it is just a tiny bit more charitable to attribute what she is saying to her bad day at work, charity instructs you to do that.

To see this, consider two rules that Jones might follow for getting himself to school:

Default to Elm Street: Take Elm Street, unless there is construction or a parade, in which case go a different way.
Go the fastest way: Take whatever street will get you there fastest.

If Elm Street tends to be the fastest, then these rules will often give the same answer, but they are not the same. Someone who defaults to Elm Street needs first to be aware of some good reason R which makes it difficult to take Elm Street, and thus provides a compelling reason to change course. Similarly, someone who relies on Default to Trust needs to notice some good reason R in order to seek a causal explanation. If this view is correct, then sometimes there will be a more charitable interpretation of the person's behavior that is *not* the interpretation you should give, because there is no good reason R which overrides the default. If you are using charity to determine whether to believe what someone says, then you do not need a reason R to rear its head: you always use charity to figure out how to understand another person's behavior.

Thus, Default to Trust is different from the norm that says to use charity in order to determine whether to take the participant stance, even if the two norms often give the same result.

Now the upshot so far is just that, if you are using charity as your interpretive norm, then you aren't taking a speaker's behavior at face value as the default. But, one might wonder, is that really a problem? Maybe we should just think: so much the worse for Default to Trust! But I actually think Default to Trust is a better principle, and that there is a good explanation for this. The explanation relies on another point from the literature on testimony. Consider the question: does testimony that p justify the belief that p because it provides evidence that p, or is the relationship between a speaker's testimony and the content that she asserts typically non-evidential? Compare:

(i)   When I yawn repeatedly during our conversation that is evidence that I am sleepy.
(ii) When I tell you that I am sleepy during our conversation that is evidence that I am sleepy.

Plausibly both are true: my telling you that I am tired is evidence, just like my yawning is. Perhaps my telling you is *better* evidence than my yawning, but, if so, that is a difference in degree, not a difference in kind.

But as Grice points out in "Meaning" (1957), there is a difference between getting someone to think that p and telling them that p. According to proponents of the *interpersonal view of testimony*, when a speaker tells you something she is not merely bringing into existence a piece of evidence as to her inner life, as she does when she yawns. Testimony provides reason for belief that is "categorically different" from that provided by evidence.[9] What is special about *telling* someone something is that it involves offering an assurance (Moran 2005), or inviting the listener to trust you (Hinchman 2005). Moran illustrates this idea by explicitly contrasting the position of Quine's "radical interpreter," on the one hand, with the position of an ordinary listener who believes a testifier, on the other. The radical interpreter observes the verbal behavior of an exotic community and takes this as evidence as to what the members of the community believe. She relies on the principle of charity to interpret their language and attribute beliefs to them. But Moran thinks that, in doing this, the radical interpreter is engaging in an activity that is very different from the activity of a person in an ordinary conversation:

> . . . one's relation to the exotic speech community does not involve being told anything at all, or believing them . . . it is just a matter of an inference from behavior which is seen as rational to some conclusion about the state of the world. So nothing along these lines, justifying the beliefs we acquire from other people, can count as a vindication of our reliance on testimony, since it is not a vindication of what we learn through believing other people.   (Moran 2005: 276)

Now, Moran and Hinchman defend the interpersonal view of testimony as an answer to the question of why we are epistemically entitled to rely on testimony to the extent that we are.[10] They think that if testimony were *merely* evidence, then we would not be entitled to rely on it to the extent that we do. There are serious objections to this proposal.[11] But again, our question is not about our epistemic entitlement to rely on testimony but, rather, about how testimony should be treated, as a normative ethical matter. We can bracket the question of whether the speaker's assurance confers epistemic justification on recipients of testimony, and take Moran's

[9] Marušić and White (2018: 105) usefully connect this distinction to the distinction between treating someone as an informant and treating them as a source of information, which Miranda Fricker (Fricker 2007: 132) borrows from Edward Craig (Craig 1990: 35) in order to explain certain cases of epistemic injustice.

[10] Other theorists who emphasize that testimony in interpersonal relationships should be understood as involving trust rather than evidence include E. Fricker (Fricker 2006) and Marušić (2015). For a critique, see Kornblith (forthcoming).

[11] See, for example, the "eavesdropper" case from Lackey (2008: 233): plausibly, if I confide in you a juicy piece of gossip, an eavesdropping colleague of ours would be as justified as you are in believing it, provided all else is equal. But I only offered *you* my personal assurance.

contrast as offering insight into our normative question. If we use the principle of charity to interpret interlocutors, as opposed to following a norm according to which taking a speaker at her word is a strong default, we fail to properly respond to the distinctive way that, in telling you something, a speaker is offering you her assurance. If in telling you something a speaker is offering you her assurance, then it can be disrespectful for you to override this assurance and speculate about non-rational causes. So that is why I think that Default to Trust is true.

5. Causal Interpretations as Debunking Explanations

Let us take stock. I have argued that, as a normative ethical matter, Default to Trust does a better job than charity at guiding our interpretations of others. But one might worry that Default to Trust doesn't really answer the question with which we began, because Default to Trust leaves it entirely open what would count as a good reason to depart from the default. Clearly, the reason must be objectively, rather than merely subjectively, good. However, even if we help ourselves to the notion of an objectively good reason, Default to Trust might still seem not to offer enough guidance. It does not exactly provide us with a mechanical procedure for determining when the default is undermined. But I believe that this is as it should be. Suppose that we set out to make a list of objectively good reasons to doubt someone. Suppose we write: *she is in a bad mood* and *she had a bad day at work* on our list. The trouble is that people often *do* say true and important things that they really mean when, and sometimes because, they are in a bad mood or had a bad day at work. And this will hold for any condition that we put on the list: it will always be possible that, although the condition is met, it has not succeeded in undermining your interlocutor's rational powers.

Part of the issue is that it is not obvious which kinds of causal influence would tend, in general, to render a piece of behavior fit for "merely causal" interpretation. An obvious suggestion is that the relevant kind of causal influence is that which undermines or supplants whatever justifying or rationalizing explanations the behavior might have had. In other words, to give a "merely causal" explanation is to give a kind of *debunking* explanation. The explanation that cites a bad day at work is supposed to undermine or supplant an explanation that cites justifying or rationalizing reasons for the behavior, and that's why it is supposed to render the behavior unfit for participant attitudes.

But one thing that emerges from the substantial literature on debunking explanations of *belief* is that there is no easy characterization of the kinds of causal information that confer debunking force.[12] From the third personal perspective, a debunking explanation of why somebody believes as she does is an explanation that either undermines or blocks her justification for believing it: something about the causal history of how she came to believe that p that should, if she is reasonable, undermine her conviction. From a first personal perspective, the realization that had you been raised by a different family then you would have very different moral or political convictions, or that had you gone to a different graduate school you would have very different philosophical views, feels unsettling, even though this counterfactual fact about your moral or philosophical upbringing doesn't bear on the question of which moral or political claims are true.[13]

On the one hand, for any belief that you hold we can point to countless causes that figure in its history that do not rationalize your believing it. Perfectly respectable etiological histories will involve countless causal forces that are irrelevant to whether the belief is true. You might possess an excellent argument for p that you found rolled up in a bottle that washed up on a beach that you would never have strolled had the coin you flipped in deciding whether to go to the beach come up heads. Intuitively, since these causes of your coming to believe that p do not bear on the question of whether p, they are simply irrelevant to whether you should believe p. But, on the other hand, certain kinds of non-rationalizing causal influence do have an unsettling effect. That you would have believed otherwise had you been raised by conservatives rather than liberals or had you gone to Harvard rather than Oxford can feel undermining even if these contingencies have nothing to do with the question of whether p is true.[14] The fact that there

---

[12] In contrast to this kind of case in which an individual becomes worried about a particular belief that she holds because of the impact of a causal factor, there are debunking arguments that target a whole range of beliefs that some group of people holds. The classic debunking arguments of Nietzsche, Marx, and Freud, and the more recent debunking arguments of Richard Joyce, Peter Singer, and Sharon Street, are of this type. These theorists offer evolutionary, cultural, or biological explanations of all or some portion of our moral beliefs. While the arguments differ in detail and in the range of beliefs that they target, they have in common the basic idea that the causal explanation for why we hold the beliefs we do does not *rationalize* those beliefs.

[13] The example of how one's choice of where to go to graduate school plays a role in shaping one's philosophical convictions is due to G. A. Cohen (2000). Cohen's example brings out that it is not only what one believes but also the reasons that one finds compelling that can be shaped by seemingly irrelevant factors.

[14] For a discussion of the epistemological significance of contingencies in the history of one's moral and political beliefs see Sher (2001).

are "mere" causes in the history of your coming to believe that p is not itself undermining, but there is no mechanical procedure for judging what kind of causal information does or should have debunking force. As the thousands of pages of discussion in both metaethics and epistemology attest, it is much easier to get a rough feel for the kinds of causal influence that are undermining than to offer a precise account.[15]

Returning now to our issue of when to view someone's behavior as "merely caused": very often, to view behavior as merely caused and thus unfit for participant attitudes is to attempt to debunk it. What debunkings of belief and behavior have in common—the feature that makes it appropriate to apply the "debunking" label to both—is the idea that the causes of a belief or a piece of behavior somehow fail to properly rationalize it. But just as the presence of non-rationalizing merely causal influence is not sufficient to undermine the justification for belief, so the mere fact there are features of a piece of behavior's history that cause but don't rationalize it is not sufficient to undermine the justification for behavior.[16] Offhand, the question of which kinds of causal influence are in fact debunking doesn't seem to have anything to do with charity. But framing the issue in terms of debunking may strengthen the case for Default to Trust. Even if there is no principled way of sorting the undermining from non-undermining "mere causes," this at least seems clear: the case for a debunking explanation is strengthened when the rationalizing explanation is *unavailable*. Thus Street's influential evolutionary debunking argument is naturally understood as involving *both* the claim that there is a compelling evolutionary explanation of our moral beliefs *and* the claim that no compelling explanation involving mind-independent moral reality is possible. This is just an instance of a more general phenomenon: the credibility of one explanatory hypothesis depends in part on the relative credibility of alternative hypotheses. So while Default to Trust offers us no mechanical procedure, nor a detailed list of practical advice, my hope is that the comparison with etiological debunking might provide us a with a general sense of why this is the case.

---

[15] For discussion, see, for example, White (2010), Dworkin (2013: ch. 4), Setiya (2013: ch. 2), and Vavova (2014).

[16] White (2010) questions whether the presence of merely causal influences have any epistemic significance at all. In response to evolutionary debunking arguments, White argues that the fact that a belief was *selected for* does nothing to undermine it. Consider the case of *Adam's party*: Adam, the host, is asking each guest whether she believes that p at the door. Adam is prepared to shoot people who say that they do not believe that p before they enter and put their bodies in the basement (p. 12). In this case, the belief that p is selected for, and everyone at the party believes it. But this doesn't undercut the justification that the partiers have for their belief that p: presumably before they entered the apartment they were either justified, or they weren't.

6. Conclusion

This chapter has addressed the question: when should I interpret someone's behavior through a causal lens, and in that way treat a person as a mere thing? According to Schroeder, charity is the appropriate principle. I have argued that it is not the appropriate principle, because it can be paternalistic to interpret someone charitably. I argued that you should take what someone says at face value unless there is good reason not to, because, in telling you something, a person is doing something like offering you her assurance. But I believe that there is no mechanical procedure for determining when a reason counts as good enough to override the default.

References

Burge, T. 1993. "Content Preservation." *Philosophical Review* 102: 457–88.

Coady, C. A. J. 1992. *Testimony: A Philosophical Study.* Oxford: Oxford University Press.

Cohen, G. A. 2001. "If you're and egalitarian, why are you so rich?" Cambridge: Harvard University Press.

Craig, Edward. 1990. *Knowledge and the State of Nature: An Essay in Conceptual Synthesis.* Oxford: Clarendon Press.

Dworkin, Ronald. 2013. *Justice for Hedgehogs.* Repr. edn. Cambridge, Mass.: Belknap Press of Harvard University Press.

Flowerree, A. K. "When to Psychologize." Forthcoming. https://akflowerree. weebly.com/work-in-progress.html.

Foley, Richard. 1994. "Egoism in Epistemology", in Schmitt (ed.), *Socializing Epistemology: The Social Dimensions of Knowledge.* Lanham, MD: Rowman & Littlefield: 53–73.

Fricker, Elizabeth. 2006. "Second-Hand Knowledge*." *Philosophy and Phenomenological Research* 73 (3): 592–618. https://doi.org/10.1111/j.1933-1592.2006.tb00550.x.

Fricker, Miranda. 2007. *Epistemic Injustice: Power and the Ethics of Knowing.* Oxford: Oxford University Press.

Hinchman, Edward S. 2005. "Telling as Inviting to Trust." *Philosophy and Phenomenological Research* 70 (3): 562–87. https://doi.org/10.1111/j.1933-1592.2005.tb00415.x.

Kornblith, Hilary. "Against Strawsonian Epistemology." Forthcomnig in Nathan Ballantyne and David Dunning, eds. *Reason, Bias, and Inquiry*. Oxford: OUP.

Lackey, Jennifer. 2008. *Learning from Words: Testimony as a Source of Knowledge. Learning from Words.* Oxford: Oxford University Press.

https://oxford.universitypressscholarship.com/view/10.1093/acprof:oso/9780
199219162.001.0001/acprof-9780199219162.

Langton, Rae. 1992. "Duty and Desolation." *Philosophy* 67 (262): 481–505.

Lewis, David K. 1983. *Philosophical Papers: Volume I.* 1st edn. New York: Oxford
University Press.

Marušić, Berislav. 2015. *Evidence and Agency: Norms of Belief for Promising and
Resolving.* Oxford: Oxford University Press.

Marušic, Berislav, and Stephen White. 2018. "How Can Beliefs Wrong?:
A Strawsonian Epistemology." *Philosophical Topics* 46 (1): 97–114. https://
doi.org/10.5840/philtopics20184616.

Moran, Richard A. 2005. "Getting Told and Being Believed." *Philosopher's
Imprint* 5 (5): 1–29. https://dash.harvard.edu/handle/1/10121963.

Plantinga, Alvin. 1993. *Warrant and Proper Function.* Oxford: Oxford
University Press.

Rawls, John. 1974. "The Independence of Moral Theory." *Proceedings and
Addresses of the American Philosophical Association* 48: 5–22. https://doi.
org/10.2307/3129858.

Schroeder, Mark. 2019. "Persons as Things." In *Oxford Studies in Normative
Ethics Volume 9.* Oxford: Oxford University Press. https://oxford.uni
versitypressscholarship.com/view/10.1093/oso/9780198846253.001.0001/oso-
9780198846253-chapter-5. Accessed March 26, 2021.

Setiya, Kieran. 2013. *Knowing Right From Wrong.* 1st edn. Oxford: Oxford
University Press.

Sher, George. 2001. "But I Could Be Wrong." *Social Philosophy and Policy* 18 (2):
64–78. https://doi.org/10.1017/S0265052500002909.

Shiffrin, Seana Valentine. 2000. "Paternalism, Unconscionability Doctrine, and
Accommodation." *Philosophy and Public Affairs* vol 2 no. 3: 205–50.

Strawson, P. F. 1962. "Freedom and Resentment (Philosophical Lecture): Read 9
May 1962." *Proceedings of the British Academy* 48 (January): 187.

Strawson, P. F. 1994. "Knowing from Words", in Matilal and Chakrabarti (eds.),
Knowing from Words. Dordrecht: Kluwer Academic Publishers.

Vavova, Katia. 2014. "Debunking Evolutionary Debunking." *Oxford Studies in
Metaethics* 9: 76–101. https://doi.org/10.1093/acprof:oso/9780198709299.003.
0004.

Webb, Mark Owen. 1993. "Why I know about as Much as You Do: A Reply to
Hardwig." *Journal of Philosophy* 90: 260–70.

White, Roger. 2010. "You Just Believe That Because . . . 1." *Philosophical
Perspectives* 24 (1): 573–615. https://doi.org/10.1111/j.1520-8583.2010.00204.x.

12

The Consequentializing Argument Against ... Consequentializing?

Paul Hurley

I: Introduction

The strategy of consequentializing features that are intuitively relevant to the deontic evaluation of actions by building them into the telic evaluation of outcomes is almost as old as consequentialism itself. But the expansion of consequentialism in recent decades beyond rankings of outcomes as better for me and better overall to outcomes better relative to me[1] has unleashed a spate of arguments for the conclusion that any ethical theory, or at least any minimally plausible candidate, can readily be converted into consequentialized form.[2] The resulting consequentializing argument for consequentialism in its most ambitious form purports not simply to win a significant battle against other ethical theories but to end the war entirely. If all plausible alternative theories can be captured without distortion in consequentialized form, and in particular if there are reasons to prefer theories in such a form, won't it be most illuminating to compare and contrast ethical theories in their consequentialized form; indeed, aren't all plausible alternatives best understood as forms of consequentialism? Even critics who answer this last question in the negative, however, typically grant that all plausible alternatives can be converted into consequentialized form.

But they cannot. Nor is the point that certain carefully configured alternatives resist consequentializing (e.g. Brown 2011). Even many standard alternatives, in particular Aristotelian virtue ethics and Kantian ethics,

---

[1] Advocates of the move to evaluator-relative ranking of outcomes include Sen (1983, 113–32) Dreier (1993, 22–40), Smith (2003, 576–98), Louise (2004, 518–36), and Portmore (2007, 2009, 2011).
[2] For examples of such consequentializing arguments, see Dreier (1993, 2011), Louise (2004), Portmore (2007, 2009, 2011), and Peterson (2010).

Paul Hurley, *The Consequentializing Argument Against . . . Consequentializing?* In: *Oxford Studies in Normative Ethics, Volume 12.* Edited by: Mark Timmons, Oxford University Press. © Paul Hurley 2022.
DOI: 10.1093/oso/9780192868886.003.0013

cannot be consequentialized. To focus the issue I adopt a standard account of consequentializing, upon which the consequentialized form of any ethical theory is "*a substantive version of consequentialism* that yields... *the same set of deontic verdicts* that it yields" (Portmore 2009, emphasis mine). The two central elements of the successful consequentializing of any target theory are that the counterpart 1) must be a substantive version of consequentialism, and 2) must satisfy deontic equivalence, yielding deontically equivalent verdicts. In what follows I will argue that the consequentializing strategy, even when applied to these standard alternatives, fails on its own terms to produce consequentialized counterparts: It can either produce a substantive version of consequentialism or secure deontic equivalence with the target theory, but not both. Moreover, I will show that the source of this failure is instructive, explaining why such alternatives are not forms of consequentialism, and why the framing of the argument invites the mistaken conclusion that they are.[3]

The argument proceeds in two stages. The first assumes what I will characterize as a *reason-independent view* of the deontic evaluation of actions.[4] A reason-independent view determines the relevant deontic values of actions through appeal to the reasons there are for the agent to perform them, but *independently of whether the agent performs the actions for these reasons*. On such a reason-independent view, if there are decisive reasons to pay a barista for a cup of coffee, then because the virtuous agent and Parfit's infamously vicious gangster (2011a, 216, 231–2) both intentionally pay the barista, their actions will receive the same deontic evaluation, even though only one of them performs the action *for those reasons*. The second stage abandons the reason-independence assumption, allowing in addition that alternative theories provide *reason-dependent* deontic evaluations of actions. Such reason-dependent deontic evaluations of actions take the agent's reasons for performing the action to be essential to at least some fundamental deontic evaluations.[5] On such views, although the virtuous person and

---

[3] Some consequentializers suggest only that there are certain pragmatic advantages in certain circumscribed contexts to working with consequentialized forms of alternative theories, but they make no claims that the theories are forms of consequentialism or that the result is deontically equivalent to the target theory. My quarrel here is not with such 'pragmatic' consequentializers, but see Schroeder (2017) for concerns about such merely pragmatic consequentializing.

[4] I take this language of reason dependence and independence from Phillip Pettit (2015), although I do not use it to mark quite the same distinction.

[5] This distinction between reason-independent deontic evaluations and reason-dependent evaluations plays a central role in ethical theory from the Greeks through the moderns (see Section II), and continues to play such a role. See, for example, Hanser's distinction between deontic evaluations that essentially appeal to "an agent's reasons for acting" and those that are

the gangster both intentionally pay for their coffee, and there are decisive reasons for each to do so, only the virtuous person performs an action for those reasons, hence only her action merits the relevant reason-dependent deontic evaluation.

I will demonstrate in the next section that both Aristotelian virtue ethics and Kantian ethics incorporate a central role for reason-dependent deontic evaluations of actions. Consequentialist ethical theories typically do not. If the consequentializing strategy assumes that the relevant deontic evaluations of target theories are all reason-independent (Stage 1), it only captures the features relevant to such reason-independent evaluations in the resulting rankings of outcomes. Implementation of the strategy upon such alternatives will fail to achieve deontic equivalence because it will elide from view all of the reason-dependent deontic verdicts central to these target theories. The result is a theory in consequentialized form, but it is not a consequentialized form of the target theory. I demonstrate in Section III that if the strategy instead recognizes that such theories provide reason-dependent deontic evaluation (Stage 2), and builds the features relevant to such evaluations into the counterpart ranking of outcomes, the resulting ranking does achieve deontic equivalence with the target theory, but it fails to yield a substantive version of consequentialism; indeed, it merely reproduces the target non-consequentialist theory. Thus, my argument will show that either the consequentializing strategy does not produce a consequentialized form of the target theory (reason-independence assumption), failing the test of deontic equivalence, or it does yield deontic equivalence (eschewing the reason-independence assumption), but only because the product is not a substantive version of consequentialism.

The consequentializing strategy cannot do what it purports to—put even standard alternatives to consequentialist ethical theories into consequentialized form. But deploying the strategy does illuminate a central distinction between these alternative theories and forms of consequentialism, and suggests why the distinction is obscured in the current debate. I will close (Section IV) by highlighting these positive insights that flow from the failure of consequentializing.

"independent of any agent's actual grounds for acting in that way" (2005, 443), and Scanlon's distinction between evaluations that appropriately appeal to "an agent's reasons for acting," (2008, 37) and those that appeal to "what reasons there are" for acting (2008, 100).

II: Stage 1–Reason-Independent Consequentializing

Within the context of the assumption that all deontic evaluation is reason-independent, both the consequentializing strategy itself and the test of deontic equivalence can seem straightforward. Such reason independence is a natural assumption for the consequentialist, for whom deontic evaluations of actions are typically taken to be a function of the value of the outcomes that they bring about and the reasons there are to bring about such outcomes, not of the reasons for which they are undertaken, but it is also assumed by many of the most influential critics of consequentialism (Thompson 1986, 1999; Scanlon 2008, 21 and 28). Within the context of the assumption that deontic statuses are all determined reason-independently, the strategy need only identify the reasons there are for determining the deontic statuses of actions on the target theory and the actions that these reasons identify as right, wrong, etc., and take these also to determine a substantive ranking of outcomes. The results are taken to support the deontic equivalence thesis by demonstrating that "for any remotely plausible non-consequentialist theory, there is a consequentialist counterpart theory that is deontically equivalent to it" (Portmore 2007, 39–40). If, for example, the target theory identifies decisive reasons deontically requiring the agent not to commit murder even to prevent two others from murdering, the consequentialized counterpart will rank higher "the outcome in which two others commit murder to the one in which she herself commits murder" (Portmore 2009, 329). The deontically required act on the target theory, not murdering, is the act that brings about the highest-ranked outcome as identified by the counterpart form of consequentialism, the outcome that the agent has the most reason to bring about (her not murdering). The consequentialized counterpart theory deontically requires the agent to perform the act that the target theory also requires the agent to perform, resulting in a deontic equivalence.[6]

Kantian ethics requires us to keep a promise even when breaking it would prevent two promise-breakings by others, and Aristotelian ethics requires us to stand by our friend even though as a result three other people will betray theirs. If the deontic evaluations involved were only reason-independent, the consequentialist counterparts to such theories would rank highest the

---

[6] Crucially, outcomes are understood on such a strategy quite broadly, such that they include the act performed. The relevant outcome of keeping my promise can thus simply be that my keeping of my promise happens.

outcomes constituted (at least in part) by the agent keeping her promise and standing by her friend. Am I morally required to keep my promise and to stand by my friend? I need only consult the relevant rankings of outcomes,[7] hence the target theories have been converted into consequentialized form.

But the reason-independent assumption is false for these standard alternatives. They both endorse reason-dependent accounts of certain deontic evaluations of actions; indeed, such reason-dependent deontic evaluations play a fundamental role on such theories. Although the issue of whether either the Aristotelian or the Kantian account also incorporates reason-independent deontic evaluations of actions is itself fraught, I grant going forward that there is a sense in which each does. The point is, rather, that each also incorporates reason-dependent deontic evaluations of actions that are irreducible to reason-independent counterparts or to other non-deontic evaluations. Because such deontic evaluations of actions are on these accounts reason-dependent, but the relevant reason-dependent features are not built into the rankings of outcomes on their consequentialized counterparts, such counterparts fail systematically to capture equivalents to these reason-dependent deontic evaluations.

I will briefly elaborate upon each of these claims, demonstrating first that reason-dependent features are central to evaluations of actions on the target theories, at least on standard interpretations. Second, I will draw upon a parallel between epistemic evaluation of beliefs and deontic evaluation of actions to make the case that on such alternative theories these are evaluations of *actions*, they are *deontic* evaluations of actions, and they are *fundamental* deontic evaluations of actions. Once these reason-dependent evaluations and the features relevant to them are taken into account, it becomes clear that on the reason independence assumption the counterpart rankings of outcomes fail to yield verdicts deontically equivalent to those of the target theories, hence that the consequentializing strategy fails to satisfy its own standards for success.

I will only say enough here to make the case that on plausible interpretations of our traditional alternatives they incorporate reason-dependent deontic evaluations of actions. Aristotelian virtue ethics takes eudaimonia, human excellence or flourishing, to be intrinsically good, and to be

---

[7] Arguments stressing the explanatory distortion and impoverishment of consequentialized forms of non-consequentialist theories are provided by Hurley (2013; 2020), Tenenbaum (2014), Schroeder (2017, 1478–82), Bauman (2019), and Sauer (2019). I set such legitimate concerns aside here in order to focus upon the more fundamental objection that standard target theories cannot be converted into deontically equivalent forms of consequentialism at all.

constituted by other intrinsically good things, e.g. virtuous traits of character and excellent relationships (friendship).[8] These valuable things are reflected in reasons for acting, including for habituating ourselves in certain ways such that the right reasons for acting in any given situation become clear to us, and we act for these reasons.

Although the claim that Aristotelian virtue ethics even engages in deontic evaluation is fraught, I will proceed on the understanding, endorsed by Annas and Anscombe, that there is a place in Aristotelian virtue ethics for right actions as "the general, vague idea of actions that you should do" (Annas 2014, 18). Let us simply grant that such virtue ethics engages in deontic evaluation of actions; moreover, that there is a role for reason-independent deontic evaluations on such an account. Such a position seems to be supported by Aristotle's own account of just and (by extension) right action:

> Acts are called just or temperate when they are the sort a just or temperate person would do.   (1105b7 II, 4)

It is not necessary to perform the action for the reasons that it is required in order to perform the just or right action, the action that the agent should perform; it is enough that the agent performs an act of the same type that the just and virtuous person would perform for such reasons.

But along with these reason-independent deontic evaluations of just and right action, the Aristotelian also offers deontic evaluations of whether or not in performing some action the agent acts justly or rightly. In particular, Aristotle is clear that for actions that are just or right in addition "to be done …justly" (1105a30) or rightly, "the agent must also be in the right state when he does them" (1105a32):

> First, he must know [that he is doing virtuous actions]; second, he must decide on them, and decide on them for themselves; and, third, he must do them from a firm and unchanging state.   (1105a 31–5; see also 1120a24–9)

The features of actions that virtue ethicists take to be relevant to their being instances of acting justly, or rightly, or as the agent ought essentially include

---

[8] There are many strains of virtue ethics, including consequentialist strains. My focus here is upon distinctively Aristotelian strains, such as Aristotle's own (1999), and those articulated more recently by Anscombe (1999) and Annas (2011).

the reasons for which they are done—that they are done for good reasons, the reasons for which a virtuous agent will perform them. That the agent performs the right action (reason-independent), and that the agent, in performing the just or right action, acts rightly or justly (reason-dependent), both utilize the same deontic terms to evaluate the actions performed. The former is a reason-independent evaluation, the latter a reason-dependent evaluation. The reason-independent evaluation applies to actions of the same "sort" or type that someone acting justly or rightly would perform. If acting justly is paying for my coffee for the right reasons reflecting the relevant things of value, the relevant "sort" of action is intentionally paying for coffee, and the agent performs the just action if she intentionally performs an action of this type. Parfit's gangster thus fails spectacularly to act justly, but performs the same type of action someone acting justly would perform, intentionally paying for his coffee, thereby performing a just action. It is acting rightly or justly that appears to be the fundamental evaluation, and right or just action that is identified derivatively, as action of the same type as acting rightly or justly; moreover, it is acting justly or rightly that is most directly evaluation of actions themselves and not of act types.[9]

The question of whether Kantian ethics incorporates reason-independent deontic evaluations is also fraught,[10] but in what follows I will grant that it does, indeed that performing the action that is in accordance with duty is performing the right action in such a reason-independent sense. But again such reason-independent deontic evaluation contrasts on the Kantian account with reason-dependent deontic evaluation, in particular with evaluation of whether or not an agent acts as duty requires ("necessitates"), out of respect for the moral law (Kant 1997, 13). Kantian ethical and moral theory takes many interrelated things, including good wills, persons valued as ends in themselves, and freedom to have intrinsic worth.[11] The value of these things is reflected in reasons to act, including reasons to structure our

---

[9] A similar distinction between wrong action and acting wrongly is highlighted by Scanlon (2008, 23 and 29), although he takes 'wrong action' to assess the action and 'acting wrongly' to assess an aspect of the agent (2008, 28). But Hanser points out that 'wrongly' clearly functions as "an adverb of manner" modifying the action, i.e. that "the object of evaluation here is an action" (2005, 274).

[10] See Scanlon (2008, 100) for a characterization of the fraught nature of this issue, and Herman (2011; 2019) for arguments challenging the assumption that for Kant deontic evaluation of actions as right and wrong is reason-independent.

[11] See, for example, Herman on Kant's account of the fundamental value of a good will (1993), and Guyer's claim that for Kant "freedom is our most fundamental value." (2000, 2)

deliberative fields through habituation (Hurley 2001; Herman 2007), and the agent acts as duty requires, out of respect for the moral law, when he performs the action for such reasons—from duty.[12] The agent who performs the act that is required by duty, but who does not act "from respect for the law," (1997, 13) fails to act as necessitated by duty—as duty requires. This would appear to be a *deontic* evaluation of the *action* as lacking what for Kant is the most salient deontic status, that it is performed as duty requires, from respect for the moral law. Duty is, for Kant, the fundamental deontic concept, and "what constitutes duty," he argues, is "the necessity of my action *from* pure respect for the practical law" (1997, 16, emphasis mine). Such deontic verdicts are on such an account fundamentally reason-dependent verdicts. The shopkeeper who fails to act from duty, merely acting in accordance with duty, does not act as duty requires him to, out of respect for the moral law. This is a negative deontic evaluation of his action; in failing to act from duty, as duty requires, he violates this fundamental deontic requirement on action. To avoid the relevant deontic censure, the action must be performed, as Herman emphasizes, not merely in accordance with duty but "from the motive of duty." (2011, 91)[13]

Such characterizations of these alternative views, and of the reason-dependent deontic evaluations that are central to them, are often dismissed either as not really, appearances notwithstanding, evaluations of actions themselves, but of the quality of will or character manifested in their performance, or as not really deontic evaluation of actions, but some other kind of evaluation, e.g. aretaic evaluations.[14] Just as evaluations of beliefs as true are fundamentally reason-independent, the suggestion is that deontic evaluations of actions are fundamentally reason-independent, and any consideration of the reasons for which the action is performed involves other non-deontic evaluation. It is evaluation of beliefs as true, reason-independent evaluation that is fundamental; similarly, it is deontic evaluation of actions as right, reason-independent evaluation that is fundamental.

---

[12] I bracket here complex questions concerning what constitutes action from duty, but see Herman's (e.g. 2007; 2019) account of the role of a Kantian agent's deliberative field in deliberation and decision-making.

[13] Mark Timmons characterizes such a reason-dependent understanding of fundamental deontic evaluation in Kant as the "motive content thesis," and presents (2002) Herman's arguments for and his own arguments against interpreting Kant as defending such a thesis. For my purposes it is only necessary to establish that it is plausible to interpret Kant as defending such a thesis.

[14] I am indebted to an anonymous referee for pressing this challenge.

To see why such a response misfires—indeed, why it falls prey to the tendency that Julia Annas warns against, of applying "theory-laden uses" of the relevant evaluative terms which "unsurprisingly do not fit virtue ethics" (2014, 15)—it is useful to explore briefly the parallel between epistemic evaluation of beliefs and deontic evaluation of actions.[15] Evaluation of a belief as true is a reason-independent evaluation. If my belief is true, where our evidence for such a judgment is that there are compelling reasons to hold it, it is true whether or not I hold it for such compelling reasons. Similarly, if my action is right, where our evidence for such a judgment is that there are decisive reasons to perform it, it is the right action whether or not I perform it for such reasons.

But the fundamental form of epistemic evaluation of beliefs is whether an agent who believes that p *knows* that p, and such evaluations, at least on standard approaches, essentially involve reason-dependent elements. The classical account of knowledge is the tripartite account, according to which knowledge is 1) justified 2) true 3) belief, and justification, the reasons for which we believe, alters in important respects what we believe. Robert Audi notes that on such accounts beliefs "can differ both in their content and in their basis" (2010, 182). In each case "we believe one thing on the basis of, so in a sense *through*, believing another . . . beliefs are mediated by other beliefs" (2010, 181). Two beliefs with the same propositional contents are in important respects different beliefs if they are based upon, hence mediated by, different beliefs. To know that p, then, the propositional content of the belief must be true, and the belief must be mediated by beliefs that provide the right reasons for holding it. If I believe for the wrong reasons, if "a crucial premise of my inference . . . is one I am unjustified in believing," then not only does my belief lack justification, but it is not the same belief as that held by one who knows that p (2010, 183); it is mediated by a flawed inferential base as the belief of the agent who knows that p is not. Such a true belief that p, mediated by an unjustified and false inferential base, will in turn ground unjustified and false beliefs. Tim Scanlon also emphasizes this aspect of belief:

> . . . accepting a reason for or against one belief affects not only that belief, but also other beliefs and the status of other reasons. This can happen in many ways. (Scanlon 1998, 52)

---

[15] I take central elements of this analogy between practical and theoretical evaluation from Herman (2011, and correspondence), although I have developed it in a somewhat different direction.

Our unjustified true believer may hold a belief with the same propositional content as one who knows that p, but he does not believe for the right reasons, hence his belief differs importantly from that of one who knows that p. One agent knows that p, the other merely believes truly that p, lacking knowledge that p; the agent's belief and his justification for holding it are flawed in the latter case, resulting in a belief that p that falls short of knowledge that p.

Consider now the parallel with actions, which, as Scanlon points out, "exhibit a similarly complex structure" (1998, 52). Two actions of the same type, e.g. two intentional actions of paying for coffee, are in important respects different actions if their performance is undertaken for different reasons reflecting different values. Such actions done for different reasons have different inferential bases, and are in this respect different actions. Like beliefs, such actions are mediated by the agent's reasons for performing them, and two very different actions in this respect may nonetheless be actions of the same type.

Such a parallel between beliefs and actions suggests a parallel between the two epistemic evaluations of beliefs, as true and as knowledge, and two corresponding deontic evaluations of actions. Just as, in the case of belief, there are fundamental reason-dependent epistemic evaluations of beliefs, whether the agent knows that p, and there are reason- independent evaluations of beliefs that p as true, as having the same propositional content as knowing that p, so too we might expect a parallel distinction between fundamental reason-dependent deontic evaluations of actions and reason-independent evaluations of types of actions. The Kantian and the Aristotelian, we have seen, can plausibly be understood as providing just such parallel reason-dependent and reason-independent deontic evaluations, reason-dependent evaluations of particular actions as acting rightly or as duty requires, as necessitated by duty out of respect for the law, and derivative, reason-independent evaluations of actions as of the same sort that someone acting rightly or as duty requires would perform—right actions. Such a right action with an unjustified inferential base, e.g. the gangster's action of paying for his coffee, will, mediated by its flawed inferential base, ground unjustified and wrong actions as constitutive and instrumental means to performing such an action of the right type. Parfit's gangster's reasons will lead him to pickpocket the money necessary to pay for the coffee, performing a wrong action in order to perform this right action, or to failure to complete payment should circumstances change (he learns the surveillance camera is broken), abandoning the right action midway through, or to keeping excess change should he receive it, effectively not paying for the coffee at all; the virtuous agent's reasons that are constitutive of her acting rightly, by contrast, will guide her in a way that forecloses all of these courses of action: She will not pickpocket the money, she will complete the payment regardless of the functioning of the security camera, and she will not keep the excess change, insisting on completing the action mediated by reasons of justice and fair dealing.

The virtuous agent acts rightly, or as required by duty; the gangster at best merely performs an action of the right type, the type that someone acting rightly or as duty requires would perform. On such accounts, the fundamental deontic evaluation of actions, like the fundamental epistemic evaluation of beliefs on the classical account, is a reason-dependent evaluation. Such an evaluation deploys the same deontic concepts as those involved in entirely reason-independent evaluations; indeed, they are the fundamental deployments of such concepts in the evaluation of actions (not just act types) on such theories.

Crucially, the point is not that the consequentialist cannot reject the centrality of such reason-dependent deontic evaluation of actions, much as reliabilists in epistemology (Foley, 2003, 314–16) reject the centrality of reason-dependent epistemic evaluation of beliefs. The point is that this is for the consequentialist, like the reliabilist, to reject the accounts of deontic and epistemic evaluation deployed by these alternatives, and with them the reason-dependent deontic and epistemic evaluations of actions that such accounts generate. The consequentialist's focus on reason-independent deontic evaluation of act types as the fundamental form of deontic evaluation departs from traditional reason-dependent approaches to deontic evaluation of actions,[16] much as the focus by reliabilists in epistemology on reliably tracking reason-independent truth as the fundamental category for epistemic evaluation is a radical departure from standard reason-dependent epistemic evaluation.

It is thus reason-dependent evaluation of beliefs as cases of knowing that p that provides the relevant parallel to reason-dependent deontic evaluation of actions as cases of acting rightly or as duty requires. Such reason-dependent evaluations are fundamental to such accounts, they are paradigmatically evaluations of the actions and beliefs themselves, and they are

---

[16] See Wiggins's arguments (2006, 139 ff.) that Aristotle, Hume, and Kant all take reason-dependent deontic evaluation to be fundamental, and that consequentialism is the outlier in this respect.

deontic and epistemic evaluations of such actions and beliefs. Reason-independent deontic evaluations of types of actions are on such alternative theories parasitic on the more fundamental reason-dependent deontic evaluations, and they seem more properly understood as evaluations of types or sorts of actions rather than of particular actions, in particular of actions as being relevantly of the same sort as the actions that agents acting rightly or as required by duty would perform.

Matthew Hanser captures the rationale, implicit in this analogy between evaluations of beliefs and actions, for the centrality of such reason-dependent deontic evaluation of actions. Actions, like beliefs, are exercises of rational powers, constituted in part by the reasons for performing such intentional actions (2005, 447) and holding such beliefs. If the reasons that an agent takes to justify do not, then in taking herself to be acting rightly she is *mis*taken, and to be mistaken is to be *mis*guided, to be prone to subsequent failures in acting rightly and performing other right actions. Because her acts and beliefs are constituted in part by the wrong reasons, she may be performing the right action, but she is not acting rightly, and she may hold a true belief, but she lacks knowledge. (2005, 448) On these alternative theories it is a fundamental deontic verdict that she fails to act rightly, or as duty requires, whether or not she performs an action of the same type as someone who does act rightly or as duty requires.

Consider, with this backdrop, Portmore's (2009) proposed "Kantsequentializing" of Kantian ethics. He takes it to be the case that the Kantsequentialist and the Kantian ethicist agree that persons are intrinsic-ally, fundamentally, non-instrumentally valuable, i.e. that they share an account of the fundamental things of value that are reflected in reasons for action. The Kantian takes such values to be reflected in reasons to refrain from murdering even to prevent two others from murdering. It is contrary to duty to murder even to prevent two others from murdering. Deploying the consequentializing strategy, the Kantsequentialist takes relevant features to be reflected in the ranking of the outcome of refraining from murdering over the outcome of murdering (to prevent two murders). There are thus decisive reasons for the Kantsequentialist to refrain.

Portmore is clear, however, that the Kantsequentialist treats the Kantian target theory as a theory of objective rightness,[17] of reason-independent deontic evaluation. When an agent intentionally refrains, she performs the right action for both the Kantian and the Kantsequentialist, and, because this

[17] See Portmore (2019, 17–24) for his account of the relevant sense of objective ought, and, by extension, of objective rightness.

is stipulated to exhaust the deontic verdicts on a theory of objective rightness, deontic equivalence is taken to be achieved. But we have seen that the Kantian theory of deontic evaluation is not merely a theory of objective rightness—of reason-independent deontic evaluation. The Kantian agent only acts as duty necessitates if she acts out of respect for the moral law—from duty, not merely in accordance with it. If the agent refrains but not for the right reasons, she does not act as duty requires on the Kantian account; she only performs an action of the same type—merely in accordance with duty. This is a fundamental form of *deontic* failure for Kant, but the Kantsequentialist counterpart fails to capture either such fundamental deontic evaluations of actions or the features relevant to them in its ranking of outcomes, hence it fails to provide deontic equivalents to such deontic failures. Whereas the Kantian issues two deontic verdicts concerning the self-interested shopkeeper, that he performs the right action in accordance with duty, and that he fails to act as duty requires from respect for the practical law, Kantsequentializing only captures right action, action merely in accordance with duty.

Reason-dependent deontic evaluations of actions are central to these alternative accounts of the deontic evaluation of actions, as reason-dependent epistemic evaluations of beliefs are central to traditional accounts of knowing that p. They are not at all central to consequentialist accounts. A consequentializing strategy that fails to recognize the central role of such reason-dependent deontic evaluations, hence that fails to build the features relevant to such evaluations into the ranking of outcomes, will fail to achieve deontic equivalence between these target theories and the consequentialized counterparts that purport to deontic equivalence with them. There is on such an interpretation of the strategy no equivalence of deontic verdicts, only at most of reason-independent deontic verdicts. But this is a failure to achieve one of the two criteria for success of the consequentializing strategy, deontic equivalence. We are left within the context of the reason independence assumption not with an expansion of the consequentialist umbrella to comprehend such alternatives, but with an appreciation of its deep structural inability to do so.

III: Stage 2: Consequentializing Without the Reason Independence Constraint

We have seen that if the consequentializing strategy captures only features relevant to reason-independent deontic evaluations of actions, counterpart rankings of outcomes will systematically fail to yield verdicts deontically equivalent to our standard alternatives. The strategy fails within the context of this assumption on its own terms. But it can seem equally clear how to avoid such failure. After all, the strategy tells us to build the features relevant to deontic evaluation of actions into the consequences, and clearly among the features that our two target theories take to be relevant to deontic evaluation of action is performance of the action for good reasons reflecting the relevant things of value. If we build such reason dependence into the outcome, then what must happen, the outcome, is not only the agent's refraining from murdering, for example, but her refraining for the right reasons reflecting the relevant values.

This move to incorporating reason-dependent features, and the rankings of outcomes that result, successfully avoids failure of deontic equivalence in cases like Parfit's gangster. Because performance of the action for the right reasons reflecting the relevant things of value is now built into the outcome, and the gangster's reasons for acting are as far as possible from the right ones, his paying for his coffee will not bring about the best outcome, and the failure of deontic equivalence on the reason independence assumption will be avoided. Similarly, the agent who keeps his promise only because it happens to be in his narrow self-interest to do so does not act for the right reasons reflecting respect for persons as ends in themselves. Because the relevant outcome is the occurrence of the performance of the action for the right reasons reflecting fundamental things of value, his action will not bring about the best outcome. Perhaps, then, the failure isn't with the consequentializing strategy but with the reason-independent constraint on its exercise?

The Scylla of failure to secure deontic equivalence, however, is avoided here only at the price of falling into Charybdis. Building only features relevant to the reason-independent evaluation of actions into the outcomes fails the test of deontic equivalence, but the price of abandoning this reason independence constraint is failure of the second test of successful consequentializing—production of a substantive version of consequentialism. The occurrence of the action of the virtuous agent will be ranked highly, and that of the gangster will not, because the best 'outcome' just is the successful performance of the right action for what the target theories identify as the right reasons, reasons reflecting what the target theories identify as the fundamental things of value. For these reason-dependent theories of deontic evaluation, building the features relevant to deontic evaluation into the outcomes requires building the entire target theory into the ranking of outcomes: the best outcome *is* the successful performance of the right action for the right reasons reflecting the relevant things of value. But the consequentialized counterpart now is not a substantial consequentialist alternative; rather, it *is* the target theory, simply viewed from a different aspect. In successfully performing the right action for the right reasons I bring it about, as a constitutive consequence, that my action happens, and what makes it the best outcome is simply that it is a constitutive consequence of acting rightly or as duty requires.

For example, Aristotelian virtue ethics holds that virtuous agents have reasons to perform actions reflecting fundamental things of value. These reasons, when decisive, are the reasons for which the agent acting rightly undertakes the performance of the action. In successfully performing the action guided by her reasons for undertaking it she brings it about, as a constitutive consequence, that such a performance for such reasons reflecting such values happens. A ranking of such constitutive consequences of actions undertaken for right or wrong reasons that reflect or fail to reflect fundamental things of value is not a substantive alternative to Aristotelian virtue ethics, it *is* Aristotelian virtue ethics, albeit presented as emphasizing not the aim of acting rightly, guided by reasons, but the constitutive consequence of successfully pursuing this aim—that acting rightly happens. Moreover, the ranking of outcomes is evacuated of substance by incorporation of the very features of these target theories that are necessary to secure deontic equivalence. Building the reason-dependent features relevant to deontic evaluation of actions into the outcomes does produce a ranking with equivalent deontic verdicts, but only because it is the same deontic ranking of actions that occurs on the target theory embedded in the same explanatory rationale reflecting the same fundamental things of value, now simply viewed from the perspective of the constituent consequents of such actions, such that acting for the right reasons reflecting the relevant things of value is the best thing that can happen.

IV: Constructive Insights: Beyond Consequentializing

It is thus not the case that all alternative theories can be put in consequentialized form. Nor do we need to manufacture carefully tailored ethical theories to demonstrate such failure; two of the standard alternatives suffice.[18] The consequentializing strategy demonstrates that substantive,

---

[18] Up to this point I have followed most consequentializers in focusing primarily upon act consequentialism. It may be thought that a broadening of or a shift in focal points from acts can avoid these obstacles to consequentializing alternative theories. Such proposals raise complex

deontically equivalent counterparts of these alternative theories cannot be produced. Consequentializing does not expand the scope of consequentialism to encompass such alternative theories; it reveals that, and why, these theories limit the scope of consequentialism. In closing I want briefly to highlight certain constructive insights that flow from this generally critical argument.

The first insight concerns rationales for deontic evaluation. Many consequentialists have alleged that unless an ethical theory is a form of consequentialism it cannot provide a value-based rationale for the deontic evaluation of actions; indeed, the claim that the good must be prior to the right is often supported through such arguments.[19] The suggestion is that without a rationale appealing to better and worse outcomes to be promoted, deontic evaluations must rely on bare appeals to intuition or question-begging appeals to "obligations themselves" (Smith 2003, 587). The quest for a rationale for deontic evaluation drives us to a value-based rationale, but, the thought continues, this is to drive us to some form of consequentialism.

The process of attempting to consequentialize our standard alternatives, however, reveals not only that they do provide value-based rationales for deontic evaluation but that 1) the consequentializing strategy presupposes that they do, and that 2) appeal to such value rationales in no way drives us to consequentialism. Each of our two alternative theories identifies particular things of fundamental value, takes such things of value to be reflected in reasons for action,[20] and takes action for such reasons to determine deontic evaluations. For the Aristotelian eudaimonia, virtuous traits of character, and excellent relationships are reflected in reasons for action that determine

issues that I cannot adequately address here, but I will briefly indicate why I do not take them to avoid the objections against consequentializing developed here, at least for the alternative theories under consideration here. The difficulty is that for such theories deontic evaluation depends upon the reasons for which the agent acts and the things of value these reasons reflect, values that cannot be captured without distortion in rankings of outcomes. On such theories agents acting rightly or as duty requires must act for such reasons reflecting such outcome resistant values. This incorporation of outcome-resistant values and the reasons that reflect them as features of deontic evaluation precludes any substantive consequentialist counterpart, regardless of evaluative focal point(s).

[19] See Scheffler (1982), Kagan (1989), and Smith (2003). Smith, for example, sees "no way of analyzing the stringency of an obligation except by way of considering the amount of good that acting on that obligation will produce" (587).

[20] For accounts of this relationship between value and reasons for action see Kolodny (2011), e.g. "Of course, it is also obvious... that we see our reasons as flowing from what is valuable, or good, or worthwhile" (69), and Scheffler's assertions that to value something is to see it "as a source of reasons for action" (2011, 27; also 29). See also Parfit's extensive discussion of goodness in the "reason-implying sense" (2011a, 38; 2011b, 432).

the telic and deontic value of actions. For the Kantian freedom, goodness of wills, and respect for persons as ends are reflected in reasons for action that determine better and worse courses of action.

The theories thus offer value rationales for deontic evaluation of actions; moreover, the consequentializing strategy presupposes that they do. Kantsequentializing, for example, takes as its point of departure the Kantian appeal to the intrinsic value of persons as fundamentally, intrinsically valuable. The Kantsequentialist purports to share the Kantian's fundamental values, and to endorse the claim that they provide a rationale for deontic evaluation. What the Kantsequentialist endorses in addition, and the Kantian denies, is a constraint on the form that any such value rationale can take, an outcome-centered constraint upon the relationship between the things of value invoked and the reasons for action that reflect them:

Consequentialist Constraint: The relevance of fundamental things of value to reasons for action and deontic evaluations of actions is captured entirely, and without distortion, in relevant rankings of outcomes to be brought about.

Although our two standard alternative ethical theories are distinguished from each other by the things that they take to be of fundamental value, and forms of consequentialism are distinguished from each other by the things that they take to be of fundamental value, consequentialism itself is distinguished from our alternatives by commitment to this outcome-centered constraint on acceptable value-based rationales. The Kantsequentialist and the well-being act utilitarian could not disagree more about what the fundamental things of value are; they are both consequentialists because they take the relevance of things of value to be captured without distortion in substantive rankings of outcomes to be brought about. The Kantian and the Kantsequentialist might seem to agree completely about what the fundamental things of value are; one is and the other is not a consequentialist because the latter takes the relevance of these things of value to reasons for action to be captured entirely in substantive rankings of outcomes to be brought about and the former does not. For the Kantian many of the reasons to act reflecting the value of persons as ends are not reasons to bring about outcomes; the Kantsequentialist claims that they all must be.[21] But this is precisely why the consequentializing strategy

<hr>

[21] See Kolodny (2011) and Hurley (2019) for general arguments that many of the reasons to act that reflect our values do not appear to be reasons to promote. See also Muñoz's (2021) distinction between telic reasons and responsive reasons.

cannot capture the features the Kantian takes to be relevant to deontic evaluation in outcomes to be brought about, save by collapsing such 'outcomes' into the constituent consequents of actions evaluated through appeal to reasons that reflect things of fundamental value, the relevance of which cannot be captured without distortion in substantial rankings of outcomes to be brought about. That Kantian and Aristotelian ethical theories cannot be consequentialized demonstrates not that they reject the appeal to value rationales, but that they reject the outcome-centered constraint upon such rationales for deontic evaluation characteristic of consequentialism.

In addition, certain lessons learned from the implementation of the consequentializing strategy, and in particular from its failure to extend the consequentialist umbrella, help explain why it is so easy to make the assumption that any value rationale must be an outcome-centered rationale, hence why it is so difficult to bring into focus the central question of whether, and if so why, the relevance of fundamental things of value to reasons for action is captured without distortion in rankings of outcomes to be brought about. For example, it seems plausible to expect that an adequate ethical theory will provide an account of the relationship between actions and outcomes, and in particular of actions as bringing about outcomes. Such an adequacy constraint might seem to favor both consequentializing and consequentialism. But we can now see that such an appearance turns on an equivocation between two very different senses in which actions can be understood as bringing about outcomes, a constituent sense and a rationalizing sense.[22] The former is all that is required as a structural feature of an adequate ethical theory, and all that our standard alternatives support. It is the unwarranted slide to the latter, rationalizing sense that appears to provide support for consequentialism and consequentializing.

This distinction comes clearly into view with recognition that, even for Aristotelian and Kantian theories, successfully performing the deontically required action brings it about, as a constituent consequent of such a performance, that the action for those reasons happens. More generally, it is a constituent consequence of successfully performing any action for any reasons reflecting any values that the action for those reasons happens: In lying to further my interests, I bring it about that my lying to further my interests happens, etc.

---

[22] See my Hurley (2019) for elaboration of this distinction.

Constitutive Sense: In successfully performing the deontically recommended action for reasons, the agent brings about the outcome that her action for these reasons happens.

Because actions successfully performed for any reasons reflecting any values bring about outcomes in this constitutive sense, it is in itself completely agnostic with respect to rival ethical theories.

But implementation of the strategy also has revealed a commitment distinctive of consequentialist theories, a commitment that invokes a distinct sense of bringing about. The commitment holds that the relevance of things of value to reasons for action is captured without distortion in relevant rankings of outcomes to be brought about. Within the context of this distinctive consequentialist commitment, reasons to act will all be (albeit perhaps indirectly) reasons to bring about the outcomes that capture the relevance of things of value to reasons for action; hence in successfully performing the deontically required action an agent is always bringing about the outcome the value of which rationalizes its performance. This commitment suggests that all deontically required actions not only bring about outcomes in the first, constitutive sense, but also in a second, rationalizing sense:

Rationalizing Sense: In successfully performing the deontically recommended action for reasons, the agent brings about the outcome the value of which provides reasons for its performance.

Within the context of the consequentialist commitment that any value rationale must be outcome-centered it is difficult to avoid the result that all relevant reasons to act are reasons to bring about outcomes in this second, rationalizing sense; hence that all deontically recommended actions are rationalized through appeal to the outcomes that they bring about. For consequentialist ethical theories, in successfully performing the deontically recommended action the agent not only always brings it about that her action happens in the constitutive sense, she also always promotes some outcome (capturing relevant values) that provides the rationale for its performance.

To highlight this distinction is at the same time to flag a tendency to ride roughshod over it, to slide from the first sense, which provides no support for consequentialism against its rivals, to the second sense, which dictates the adoption of consequentialism in preference to its rivals. If, in keeping my promise, I bring it about that my promise keeping happens (constitutive sense), it can seem natural to slide to the claim that my aim, in keeping my promise, is to bring it about that my promise keeping happens (rationalizing sense). The illicit slide suggests that it is the value of the outcome, my promise keeping happening, that rationalizes my keeping my promise to bring this outcome about.[23] But this is precisely what the Aristotelian and the Kantian deny. The relevant reasons to keep my promise, they hold, reflect values that cannot be captured without profound distortion in rankings of outcomes to be brought about. In acting for the right reasons, completing the required performances of actions guided by the reasons for undertaking them, agents bring it about in the constitutive sense that their promise keepings happen for those reasons. We have seen that the attempt to capture the resulting reason-dependent deontic evaluation of such theories in rankings of outcomes leads to the collapse of rankings of outcomes into constituent consequents of deontically required actions—to a ranking of outcomes brought about only in the constitutive sense, not in the rationalizing sense.

I take no position here regarding the truth of the consequentialist constraint, hence no position on whether deontically recommended actions do always bring about outcomes not merely in the constitutive sense, but in the rationalizing sense as well.[24] My point is that our alternative theories reject this constraint, that consequentialist alternatives endorse it, and that the tendency to slide between the two senses of bringing about obscures the importance of this distinctive constraint, and the need to focus going forward upon grounds for accepting or rejecting it. Insofar as the conse-quentializing strategy, faithfully implemented, highlights this distinctive constraint, it does not show us that we are all consequentialists now; rather, it clarifies why so many of us are not, why the reasons we are not tend to be elided from view in the consequentializer's framing of the debate, and that it is the strength of the case for and against this distinctive consequentialist constraint that will determine who is right.

Bibliography

Annas, Julia (2011). *Intelligent Virtue*. Oxford: Oxford University Press.

Annas, Julia (2014). "Why Virtue Ethics Does Not Have a Problem with Right Action." In *Oxford Studies in Normative Ethics Vol. 4*, ed. Mark Timmons. Oxford: Oxford University Press.

---

[23] I have discussed this slide in Hurley (2019).
[24] I have sketched grounds for challenging this constraint elsewhere (e.g. Hurley 2018, 2019).

Anscombe, G. E. M. (1999). "Practical Truth." *Logos* 2, 68–76.

Aristotle (1999). *Nichomachean Ethics.* Trans. Terence Irwin. Indianapolis: Hackett.

Audi, Robert (2010). *Epistemology, 3rd Edition.* New York: Routledge.

Baumann, Marius (2019). "Consequentializing and Underdetermination." *Australasian Journal of Philosophy* 97, 511–27.

Brown, Campbell (2011). "Consequentialize This." *Ethics* 121, 749–71.

Dreier, Jamie (1993). "Structures of Normative Theories." *The Monist* 76, 22–40.

Dreier, Jamie (2011). "In Defense of Consequentializing." In *Oxford Studies in Normative Ethics Vol. I,* ed. Mark Timmons. New York: Oxford University Press.

Foley, Richard (2003). "Justified Belief as Responsible Belief." In *Contemporary Debates in Epistemology,* ed. M. Steup and E. Sosa. Malden: Blackwell, 313–26.

Guyer, Paul (2000). *Kant on Freedom, Law, and Happiness.* New York: Cambridge University Press.

Hanser, Matthew (2005). "Permissibility and Practical Inference." *Ethics* 115, 443–70.

Herman, Barbara (1993). "Leaving Deontology Behind." In *The Practice of Moral Judgment.* Cambridge: Harvard University Press.

Herman, Barbara. (2007). "Making Room for Character." In *Moral Literacy.* Cambridge: Harvard University Press.

Herman, Barbara (2011). "A Mismatch of Methods." In *On What Matters Vol II.* Oxford: Oxford University Press: 83–115

Herman, Barbara (2019). "Being Prepared: From Duties to Motives." In *Oxford Studies in Normative Ethics Vol 9,* ed. Mark Timmons. Oxford: Oxford University Press.

Hurley, Paul (2001). "A Kantian Account of Desire-based Justification." *Philosophers' Imprint* I, 1–16.

Hurley, Paul (2013). "Consequentializing and Deontologizing: Clogging the Consequentialist Vacuum." In *Oxford Studies in Normative Ethics Vol. 3,* ed. Mark Timmons. Oxford: Oxford University Press, 123–53.

Hurley, Paul (2018). "Consequentialism and the Standard Story of Action." *Journal of Ethics* 22, 25–44.

Hurley, Paul (2019). "Exiting the Consequentialist Circle: Two Senses of Bringing It About." *Analytic Philosophy* 60, 130–63.

Hurley, Paul (2020). "Consequentializing." In *The Oxford Handbook of Consequentialism,* ed. Douglas Portmore. New York: Oxford University Press, 25–45.

Kagan, Shelly (1989). *The Limits of Morality*. Oxford: Oxford University Press.

Kant, Immanuel (1997). *Groundwork of the Metaphysics of Morals*, ed. Mary Gregor. Cambridge: Cambridge University Press.

Kolodny, Niko (2011). "Aims as Reasons." In *Reasons and Recognition*, ed. Samuel Freeman, Rahul Kumar, and Jay Wallace, 43–78. New York: Oxford University Press.

Louise, Jennie (2004). "Relativity of Value and the Consequentialist Umbrella." *Philosophical Quarterly* 54, 518–36.

Muñoz, Daniel (2021). "The Rejection of Consequentializing." *Journal of Philosophy* 118/2, 79–96.

Parfit, Derek (2011a). *On What Matters Vol. I*. Oxford: Oxford University Press.

Parfit, Derek (2011b). *On What Matters Vol II*. Oxford: Oxford University Press.

Peterson, Martin (2010). "A Royal Road to Consequentialism?" *Ethical Theory and Moral Practice* 13, 153–69.

Pettit, Philip (2015). *The Robust Demands of the Good*. Oxford: Oxford University Press.

Portmore, Douglas (2007). "Consequentializing Moral Theories." *Pacific Philosophical Quarterly* 88, 39–73.

Portmore, Douglas (2009). "Consequentializing. "*Philosophy Compass* 4, 329–47.

Portmore, Douglas (2011). *Commonsense Consequentialism*. Oxford: Oxford University Press.

Portmore, Douglas (2019). *Opting for the Best*. New York: Oxford University Press.

Sachs, Benjamin (2010). "Consequentialism's Double-Edged Sword." *Utilitas* 22, 258–71.

Sauer, Hanno (2019). "The Cost of Consequentialization." *Metaphilosophy* 50, 100–9.

Scanlon, T. M. (1998). *What We Owe to Each Other*. Cambridge: Harvard University Press.

Scanlon, T. M. (2008). *Moral Dimensions*. Cambridge: Belknap Press/Harvard University Press.

Scheffler, Samuel (1982). *The Rejection of Consequentialism*. Oxford: Oxford University Press.

Scheffler, Samuel (2011). "Aims as Reasons." In *Reasons and Recognition*, ed. Samuel Freeman, Rahul Kumar, and Jay Wallace. New York: Oxford University Press.

Schroeder, Andrew (2017). "Consequentializing and its Consequences." *Philosophical Studies* 174, 1475–97.

Sen, Amartya (1983). "Evaluator Relativity and Consequential Evaluation." *Philosophy and Public Affairs* 12, 113–32.

Smith, Michael (2003). "Neutral and Relative Value after Moore." *Ethics* 113, 576–98.

Tenenbaum, Sergio (2014). "The Perils of Earnest Consequentializing." *Philosophy and Phenomenological Research* LXXXVIII, 233–40.

Thompson, J. J. (1986). "The Trolley Problem." In *Rights, Restitution, and Risk*, ed. William Parent. Cambridge: Harvard University Press.

Thompson, J. J. (1999). "Physician-Assisted Suicide: Two Moral Arguments." *Ethics* 109, 497–518.

Timmons, Mark (2002). "Motive and Rightness in Kant's Ethics." In *Kant's Metaphysics of Morals: Interpretative Essays*, ed. M. Timmons. Oxford: Oxford University Press.

Wiggins, David (2006). *Ethics*. Cambridge: Harvard University Press.

13
Moral Worth and Our Ultimate Moral Concerns

Douglas W. Portmore

1. Moral Worth

Some right acts have what philosophers call *moral worth*.[1] A morally worthy act manifests the agent's virtuous motives such that they deserve credit for having acted rightly.[2] And, for an agent to deserve credit for having acted rightly, not only must their act be right, but their motives must be such that their acting rightly was no mere accident. More precisely, then, a right act has moral worth if and only if it was motivated in a way that makes it non-accidentally right.[3]

As Kant noted, not all right acts have moral worth. Take his example of the shopkeeper (G 4: 397). The shopkeeper deals honestly with his customers, always giving them the correct change, but only because it's good for business. So, although he acts rightly in dealing honestly with his customers, his acts lack moral worth in that they were motivated solely out of a concern

---

[1] By 'right', I mean 'objectively morally right'. And I use the term broadly to include both obligatory acts and supererogatory acts.

[2] As I see it, to act rightly is merely to do what is right; it is not necessarily to do what is right *for the reasons that make it right* (cf. Hanser 2005).

[3] 'Moral worth' is not an ordinary language term (see Johnson King 2020, 189). It is, instead, a term of art that's most often used to describe an act as being motivated in a way that makes it non-accidentally right. To illustrate, consider the following representative quotes: (1) "what matters [for moral worth] is that the action is in accord with duty and *it is no accident that it is*" (Baron 1995, 131); (2) "when we say that an action has moral worth, we mean to indicate (at the very least) that the agent acted dutifully from an interest . . . that therefore makes its being a right action the nonaccidental effect of the agent's concern" (Herman 1981, 366); and (3) "morally worthy actions are motivated in a way that makes their rightness neither 'contingent' nor 'precarious'—they are . . . motivated in a way that makes them non-accidentally right" (Sliwa 2016, 394 and 398). See also Arpaly (2002, 225), Markovits (2010, 206 and 211), Isserow (2020, 532), Johnson King (2020, 191), Singh (2020), and Howard (2021b, 306–11).

Douglas W. Portmore, *Moral Worth and Our Ultimate Moral Concerns* In: *Oxford Studies in Normative Ethics, Volume 12.* Edited by: Mark Timmons, Oxford University Press. © Douglas W. Portmore 2022.
DOI: 10.1093/oso/9780192868886.003.0014

to maximize profits, a concern that would have led him to short-change his customers had they been less savvy. Thus, his acting rightly was merely fortuitous.

The moral worth of an act is a function of the virtuousness of the motives/ concerns it issues from as opposed to the virtuousness of the character of the agent who performs it. Thus, an act can have moral worth even if its agent has a bad character. Consider that even a stingy miser might do something generous on occasion.[4] And if, on that occasion, what moves them is an appropriate set of concerns, then their act will have moral worth even if it's out of character. Likewise, someone with a good character might act from bad motives on occasion, and, when they do, their act will lack moral worth. So, whereas the moral worth of an act depends on what the agent's motives were in the given situation and whether those motives could potentially lead them to act wrongly in other situations, an agent's character has to do with whether they're disposed to have the appropriate concerns in a wide range of situations, even if not necessarily in the given one.

As I see it, an act's having moral worth isn't the same as its agent being praiseworthy for performing it. Admittedly, an agent wouldn't deserve praise for acting rightly unless their acting rightly was non-accidental. But an agent can deserve praise for, say, *acting selflessly* even if they don't deserve praise for *acting rightly*. Thus, that an act was non-accidentally right isn't the only possible reason for praising it (see Johnson King 2020, 191). I deny, then, that "the moral worth of an action is the extent to which the agent deserves moral praise or blame for performing the action" (Arpaly 2002, 224). And, so, I will not "speak interchangeably of a morally praiseworthy action and an action which has positive moral worth" (Arpaly 2002, 224), for I see no reason to introduce a technical term as a stand-in for an ordinary one.

Also, unlike some others (e.g., Markovits 2010), I don't see moral worth as something that comes in degrees. For either an agent's motives were such that their acting rightly was non-accidental or they weren't. Nonetheless, I concede that some may use the term differently than I do. But most in the literature use the term as I do. Of course, one may still worry that as we (the majority) use it, it refers to some rather uninteresting notion. But this is not the case. For there are at least two reasons to take an interest in whether an act was motivated in a way that makes it non-accidentally right. First, as pointed out above, whether we should praise someone for acting rightly

---

[4] I'm assuming that for an act to be generous it needn't issue from a stable disposition to be generous. So, I reject what Hurka (2006) calls *the dispositional view*.

depends on whether their actions were motivated in a way that makes them non-accidentally right. Second, we're often interested in whether someone who has acted rightly in one situation can be counted on to act rightly in other situations. And that's precisely what this notion tells us.

I also differ from several others (e.g., Markovits 2010 and Sliwa 2016) in thinking that the moral worth of an act doesn't just depend on the specific concerns that moved the agent to perform it; I believe that the agent's other concerns matter as well. To see why, consider the following.

The Dog-Lover: A dog-lover named Yunn protects a poodle from a boy's kick by blocking his blow with her own leg. And she does so out of a concern for the dog's welfare. Thus, she does the right thing for the right reason. But suppose that, in this instance, Yunn had absolutely no concern for the welfare of the boy and cared only for herself and the dog. So, she would have fatally shot the boy had this been an option for her. For, in that case, she could have protected the dog without having to suffer his painful blow. But, as it was, she didn't have this option and could protect the dog only by blocking his blow with her own leg. So, her acting rightly in this instance was merely accidental. Indeed, the same set of concerns that led her to do the right thing in this situation would lead her to do the wrong thing in other relevantly similar situations, such as the one in which she has the option of shooting the boy.[5]

The lesson, I take it, is that we must look not only at the specific motive/concern that moved her to act as she did but at her other pertinent concerns as well. And I'll be assuming that a lack of concern for something counts as a concern; it just counts as a "zero-concern" for that thing. Thus, in *The Dog-Lover*, we must consider not only Yunn's concern for both herself and the dog but also her zero-concern for the boy. For she would have been led by

---

[5] Julia Markovits (2010, 210) talks about a somewhat similar case. In her case, a fanatical dog-lover saves several strangers at great risk to themself. But, given their fanatical love for dogs, they would not have saved these strangers had the choice been between saving them and saving their dog. Markovits claims that, assuming that the dog-lover's preference for saving their dog over the strangers is the result of their having too much concern for their dog rather than too little concern for the strangers, their act of saving the strangers has moral worth despite the fact that their excessive concern for their dog would have led them to do the wrong thing in other situations. Now, I'll concede this point to Markovits provided that what interests us is whether the dog-lover is willing to sacrifice their own interests for the sake of promoting the much greater interests of others. But, as I'll argue below, there may be other contexts in which what interests us is whether the dog-lover's concern for their dog is excessive. And, in those contexts, their act would not count as having moral worth on the view that I'll be defending.

this set of concerns to act wrongly in a situation in which she had the option of shooting the boy. Thus, her acting rightly in this situation was merely an accident, as it was merely fortuitous that she didn't have this other option. And, so, we should think that an act's moral worth depends not merely on the agent's motivating reason for performing it but on all their pertinent concerns.

What are the pertinent concerns? They are all and only those that will (or would) determine whether the agent acts rightly in this and other relevantly similar situations. Thus, in *The Dog-Lover*, Yunn's zero-concern for the boy's welfare is pertinent given that it's part of a set of concerns that would have led her to act wrongly in the relevantly similar situation in which she had the option of shooting the boy. Likewise, if Yunn's set of pertinent concerns had included a concern for the boy but not for dogs with spots, her acting rightly would have counted as merely accidentally right. For such a set of concerns would have led her to refrain from acting rightly (that is, to refrain from blocking the boy's blow) in the relevantly similar situation in which the boy was about to kick a Dalmatian rather than a poodle.

Of course, not every concern that determines whether an agent would act rightly in some other situation is pertinent. When it comes to pertinence, it's only the *relevantly similar* situations that matter. Thus, even if Yunn had had a concern to prevent Muslims from immigrating to the U.S., this wouldn't itself prevent her act of protecting the dog from having moral worth. Although this concern would have led her to do the wrong thing in a situation in which she had the opportunity to prevent a Muslim with a compelling asylum claim from immigrating to the U.S., this situation isn't relevantly similar to the one at hand: one in which she has to choose whether and how to protect a dog from physical violence. That said, there isn't any simple way of spelling out what the relevantly similar situations are. Thus, we may wonder whether a situation in which a boy is about to beat a snake with a stick is relevantly similar to the one described in *The Dog-Lover*. That is, we may wonder whether Yunn's having a concern for all mammals but not for any reptiles would make her acting rightly in *The Dog-Lover* count as merely accidental.

I suspect that the answers to such questions depend on the context in which they're being asked and on what's taken to be relevantly similar given our interests in those contexts. Thus, we could imagine one context in which we're interested in whether Yunn is speciesist and, thus, with whether she has a concern for the welfare of all sentient beings and not just for the welfare of her fellow mammals. In that context, a situation in which the boy is intending to beat a snake would count as relevantly similar. And, so, her acting rightly in *The Dog-Lover* would count as merely accidental. But we could also imagine a context in which we're only interested in whether Yunn is sufficiently altruistic with respect to the other members of her community and, thus, with whether she's willing to sacrifice her own welfare for theirs when appropriate. (And let's assume that her fellow humans and their canine companions, but no reptiles, count as members of her community.) In this context, the situation involving the snake wouldn't count as relevantly similar. And, so, her acting rightly in *The Dog-Lover* would count as non-accidental—remember we're assuming that, in this version of the case, she has a concern for all mammals, which includes the boy.

As I see it, this context-sensitivity is a feature rather than a bug. First, moral worth has to do with whether an agent was just lucky to have acted rightly, and, in general, whether someone counts as lucky is context sensitive. Take, for instance, the case of Kasamba. Like 90 percent of those living in his village, he contracted Ebola during a recent Congolese outbreak. Yet, unlike 75 percent of his fellow infected villagers, he survived. By comparison, far less than 1 percent of the world population ever contracts Ebola, and, of those that do, about 50 percent survive. Given all this, we may wonder whether Kasamba counts as lucky. And it seems that there is just no unequivocal answer. In a context in which our interests make everyone in the world the relevant comparison class, we should think that Kasamba was extremely unlucky to have contracted Ebola and only somewhat lucky to have survived the infection. But, if instead our interests make his fellow villagers the relevant comparison class, we should think that he wasn't at all unlucky to have contracted Ebola but was very lucky to have survived the infection.

Second, this sort of context-sensitivity explains how it's possible for philosophers to draw opposite conclusions from the very same case: the case of Huck Finn's helping his friend Jim, a runaway slave, to elude authorities despite his believing that it is wrong for him to do so. Arpaly (2002) and Markovits (2010) conclude from this case that an act can have moral worth even if it wasn't motivated out of a concern to do what's right, for, as they see it, Huck's act was non-accidentally right. But Johnson King (2020) and Sliwa (2016) conclude the opposite: that an act cannot have moral worth unless it was motivated out of a concern to do what's right, for, as they see it, Huck's act was merely accidentally right. Context-sensitivity explains this. In a context in which we're interested in whether Huck might be led to do wrong by his desire to help a friend and his willingness to do so even when he believes that it's wrong to do so, Huck's act counts as only accidentally right. For it seems that Huck just lucked out in that Jim was a fugitive slave rather than, say, a fugitive thief or murderer. But in a context in which we're interested only in whether Huck would ever be led by his appropriate concern for Jim's humanity to do wrong, his act counts as non-accidentally right, because this concern would never lead him to do wrong—at least, not when he's relevantly informed and it's combined with other appropriate concerns. So, whether we think that Huck's act was only accidentally right just depends on the context.

At this point, we should have a good enough grip on what moral worth is to be able to evaluate various proposed substantive views. I'll start by looking at two simple views in Sections 2 and 3. Although no contemporary philosopher accepts such a simple view, most contemporary views can be seen as more sophisticated versions of these two. Nonetheless, I believe that both views need more than mere revision, as I believe that both are fundamentally flawed. In Section 4, I explain what this fundamental flaw is. And this leads me to introduce a new concept—the concept of *an ultimate moral concern*— in Section 5. Lastly, in Section 6, I employ this concept in developing a new account of moral worth and show how this account compares favorably to its rivals.

2. The Simple Kantian View

Most contemporary accounts of moral worth stem from the views of either Hume or Kant. I'll present only the simplest version of each, and I make no claim as to their historical accuracy. My aim is merely to lay out the two most basic points of view from which most contemporary views have spawned. I'll start with the view that's been inspired by Kant and his thought that moral worth attaches to right actions that are performed simply because they are right—i.e., actions motivated "from duty" (G 4:396–401).

The Simple Kantian View: A right act has moral worth if and only if it was motivated out of a nonderivative desire/concern to do what's right (whatever that may be) and the belief that were they to perform this act they would do what's right.

The Simple Kantian View is subject to counterexample. Here's one.

The Empathic: A man named Christoforos denies that chimpanzees are morally considerable beings, for he mistakenly believes that they can be neither harmed nor wronged. Yet, he finds himself empathizing with the apparent plight of a chimpanzee that has just been captured by poachers. Intellectually speaking, he doesn't believe that there are any genuine feelings underlying the chimpanzee's outward "signs" of distress. But, on an emotional level, he accurately perceives that the chimpanzee is in genuine distress. And, given these perceptions, he empathizes with the chimpanzee's plight, having once been held captive himself.[6] So, when the opportunity arises, he's moved to help the chimpanzee escape back into the wild out of a nonderivative concern to alleviate what he correctly perceives to be the chimpanzee's distress.[7] And this is his sole motive, for he doesn't think that his helping the chimpanzee escape is the right thing to do. (Nor does he think that it's the wrong thing to do.) Afterwards, he's tempted to just walk away. Yet, he ends up reporting the poachers to the authorities out of both a nonderivative concern for doing what's right (whatever that may be) and the belief that reporting lawbreakers to the authorities is his duty. What's more, he facilitates the authorities taking the poachers into custody safely by hiding their guns, and he does so out of a concern for the welfare of both the poachers and the authorities. He does this knowing that it involves substantial risk to himself.[8]

On the Simple Kantian View, Christoforos's act of helping the chimpanzee escape back into the wild doesn't have moral worth because it fails to

---

[6] I believe that, through our emotional experiences, we can apprehend important truths. And these experiences provide us with evidence for these truths. What's more, they can represent the world as being one way even while our avowed beliefs represent the world as being another way. And sometimes it's our emotions rather than our avowed beliefs that accurately represent the way the world is. (See Furtak 2018, ch. 3.) That's what I take to be going on with Christoforos. Through his empathic response to the apparent signs of the chimpanzee's distress, he accurately represents the world as being one in which the chimpanzee is suffering, and yet, through his beliefs (or, at least, the propositions to which he's willing to assent), he inaccurately represents the world as being one in which the chimpanzee is not suffering. Fortunately, his actions are being guided by what his emotions are telling him rather than by what his avowed beliefs are telling him.
[7] He wants to alleviate the chimpanzee's distress for its own sake and not merely as a means to alleviating the unpleasantness that the chimpanzee's distress is causing him.
[8] More commonly, philosophers (such as Arpaly 2002) cite the case of Huck Finn as a putative counterexample to the Simple Kantian View. I prefer this example, because it's unclear whether Huck's concerns are appropriate. For instance, one might worry that Huck has too great a concern for being loyal to a friend. Consequently, Johnson King (2020, 195) worries that Huck could be led by such a concern to help Jim even if he were a fugitive murderer rather than a fugitive slave.

MORAL WORTH AND OUR ULTIMATE MORAL CONCERNS   283

manifest a nonderivative concern for doing what's right. Intuitively, though, it seems to be non-accidentally right, for it was motivated in a way that makes his acting rightly extremely reliable. After all, he had a concern not only to alleviate the chimpanzee's distress but also to safeguard the welfare of both the poachers and the authorities. He even had a nonderivative concern for doing what's right. What's more, the magnitude of each of these concerns was, we'll assume, at the appropriate level. And, given all this, he would never be led to act wrongly by such a set of concerns—at least, not if he were relevantly informed. So, we should think that, contrary to what the Simple Kantian View implies, his act of helping the chimpanzee was non-accidentally right and, consequently, of moral worth. So, contrary to the Simple Kantian View, we should deny that a right act's manifesting a nonderivative concern for doing what's right is necessary for that act to have moral worth.

Also, contrary to the Simple Kantian View, we should deny that a right act's manifesting a nonderivative concern for doing what's right is sufficient for that act to have moral worth. To illustrate, consider the following.[9]

Golfing for Rightness' Sake: Unless a patient's tumor is removed this afternoon, he'll die this evening—though not painfully. Cutting and stitching this afternoon by the only two available doctors, Slice and Patch, is the only thing that can save him. For Slice is the only one who can cut out the tumor, and Patch is the only one who can stitch him up afterwards. If there is either cutting without stitching or stitching without cutting, the patient's death will be physically agonizing. It would even be cruel for one of the two doctors to show up to the hospital knowing that the other won't, as this would only needlessly get the patient's hopes up, making his death psychologically agonizing. Unfortunately, both Slice and Patch care more about keeping their rather trivial promises to take their husbands golfing than about saving their patient's life. Consequently, each doctor is going to take her own husband golfing this afternoon regardless of what the other doctor is willing to do. And each is immovable in this regard. What's more, each knows this about the other. Thus, each knows that, given the other's unwillingness to do her part in saving the patient, taking her own husband golfing is the right thing to do. For let's assume that, given the unwillingness of the other to participate in saving the patient, the best thing that she can do

---

[9] These are adapted from David Estlund's (2017, 53) case entitled "Slice and Patch Go Golfing."

Table 1 Slice and Patch

	Slice cuts	Slice goes golfing with her husband
Patch stitches	They produce the best world that they could together produce: (O_1) the patient is saved, although both husbands are disappointed not to go golfing. Both Slice and Patch maximize utility.	They produce the world that's tied for the worst world that they could together produce: (O_2) the patient dies in agony, Patch's husband is disappointed not to go golfing, but Slice's husband is glad to go golfing. Neither Slice nor Patch maximize utility.
Patch goes golfing with her husband	They produce the world that's tied for the worst world that they could together produce: (O_3) the patient dies in agony, Slice's husband is disappointed not to go golfing, but Patch's husband is glad to go golfing. Neither Slice nor Patch maximize utility.	They produce the second-best world that they could together produce: (O_4) the patient dies painlessly and both husbands are glad to go golfing. Both Slice and Patch maximize utility.

is to take her own husband golfing as promised. Now, each has a nonder-ivative concern for doing what's right (whatever that may be). But, unfortu-nately, this desire is not as strong as each's desire to keep her promise to her husband, which is what explains why neither is willing to do her part in saving the patient. So, in the end, each doctor takes her husband golfing out of a nonderivative concern for doing what's right. Consequently, they together produce what I've labelled in Table 1 as outcome O_4—the outcome in which each goes golfing with her own husband while the patient dies painlessly.

I'm assuming that, on the correct moral theory, one is required to keep one's promises whenever there's nothing better that one can do.[10] And there is nothing better that either doctor can do. For, given Patch's unwillingness to cooperate, Slice's showing up to the hospital has no chance of saving the patient and every chance of making the patient's death psychologically agonizing. Likewise, given Slice's unwillingness to cooperate, Patch's show-ing up to the hospital has no chance of saving the patient and every chance

[10] I'm also assuming that, on the correct moral theory, it would be permissible to break one's promise to take one's spouse golfing in order to save a life.

of making the patient's death psychologically agonizing. So, given the circumstances, each doctor should stay away from the hospital and instead take her husband golfing, which is the best thing that either of them can do under the circumstances. Of course, many non-maximizing theories would deny that an agent must always perform her best option. But every plausible moral theory (maximizing or non-maximizing) will require an agent to perform her best option when doing so would fulfill a promise, maximize utility, and neither harm nor disrespect anyone. So, it seems safe to assume that, on any plausible moral theory, the right thing for each doctor to do is to take her own husband golfing.

In *Golfing for Rightness' Sake*, each doctor was motivated to take her husband golfing out of a nonderivative desire to do the right thing (whatever that may be) and the belief that if she were to take her husband golfing she would be doing the right thing. So, on the Simple Kantian View, each doctor's act of taking her husband golfing has moral worth. But, although each doctor did the right thing, neither of them deserve credit for acting rightly. For it was merely fortuitous that they acted rightly. Indeed, the same motives that led them to act rightly in this situation would have led them to act wrongly in the relevantly similar situation in which the other doctor is willing to do her part in saving the patient. So, we should reject the Simple Kantian View. Manifesting a nonderivative concern for doing what's right is insufficient to confer moral worth on an action.

3. The Simple Humean View

The other leading inspiration for accounts of moral worth is Hume. According to Hume, "no action can be virtuous, or morally good, unless there be in human nature some motive to produce it, distinct from a sense of its morality" (T 3.2.1.7). Here's a simple version of his view.

The Simple Humean View: A right act has moral worth if and only if it was motivated out of a nonderivative desire/concern to perform an act with right-making feature RMF (conceptualized as such) and the belief that were they to perform this act they would perform an act with RMF.

This view is also subject to counterexample, though what sort of counterexample depends on which of the following two versions of it we have in mind.

On the fundamentalist version, 'RMF' refers to whatever the *fundamental* right-making feature of acts is. Thus, if maximizing act-utilitarianism is correct, 'right-making feature RMF' refers to 'the feature of maximizing utility'.[11] On this view, a right act will have moral worth if and only if it was motivated out of a nonderivative desire to maximize utility and the belief that were they to perform this act they would maximize utility.

On the non-fundamentalist version, by contrast, 'RMF' can refer to any right-making feature, fundamental or non-fundamental. To illustrate, assume for the sake of argument that maximizing act-utilitarianism is correct and suppose that I would maximize utility if and only if I were to push the button that's in front of me, for pushing this button is my only option for saving many lives. Given these assumptions, the non-fundamentalist version of the Simple Humean View implies that my act of pushing the button would have moral worth if I were motivated out of a nonderivative desire to maximize utility and the belief that, were I to perform this act, I would maximize utility. But it also implies that my act of pushing the button would have moral worth if I were instead motivated to perform it out of a nonderivative desire to save many lives and the belief that, were I to perform this act, I would save many lives. For, in this instance, saving many lives is what would maximize utility. Thus, saving many lives is what makes my pushing the button the right thing to do. It's just that this is derivatively so. Thus, on the non-fundamentalist version of the Simple Humean View, 'right-making feature RMF' can refer to 'the feature of maximizing utility', 'the feature of saving many lives', or any other right-making feature.

Both versions of the Simple Humean View are problematic. The problem with the non-fundamentalist version is that it gets the wrong result in cases like *The Dog-Lover*. For this version of the Simple Humean View implies

---

[11] Arpaly and Schroeder (2014) endorse the fundamentalist version of the Simple Humean View. They hold that acting for the right reasons (which they equate with acting in a way that has moral worth) is just a matter of having intrinsic desires that are instances of good will. And they say that "for an intrinsic desire to be an instance of good or ill will the content of the desire must be something one has a *pro tanto* moral reason to do or avoid and this content must be presented by concepts that would allow the individual in question to trivially deduce that it is necessarily an instance of MAXIMIZING HAPPINESS, or RESPECTING PERSONS, or whatever the correct normative theory distinguishes as the right or good as a whole" (2014, 167). Thus, as Arpaly explains in another work, "a morally worthy action stems from a commitment to the right and the good *correctly conceptualized*. If utilitarianism has the right account of the features that make actions right then the agent performing a morally worthy action conceives of her action as maximizing utility, and is committed to maximizing utility so conceived" (2015, 87).

that Yunn's act has moral worth given both that she had a nonderivative concern to prevent the dog from getting hurt and that this is what makes her blocking the boy's blow the right thing to do. Of course, it's not what fundamentally makes it right. For assuming (merely for the sake of argument) that maximizing act-utilitarianism is correct, what fundamentally makes it right is that doing so would maximize utility.[12] Nevertheless, that the act prevents the dog from getting hurt is what derivatively makes it right given that preventing the dog from getting hurt is what would maximize utility. So, on the non-fundamentalist version, Yunn's act has moral worth. But, as we saw above, Yunn's act was merely accidentally right given that she had zero concern for the boy's welfare. Thus, the non-fundamentalist version of the Simple Humean View should be rejected.

The problem with this version of the Simple Humean View is that it fails to capture the counterfactual reliability that's required for moral worth. Consequently, it allows that the set of concerns that confers moral worth on your act is one that could lead you to act wrongly in other situations in which you are relevantly informed, making your acting rightly in this instance merely accidental. This is because the only way to ensure counterfactual reliability when relevantly informed is to consider not only the agent's motivating reason and whether it was good but also whether the agent had all the other pertinent concerns and in the appropriate proportions. This is because whether an act is permissible depends not merely on whether it has some good feature (which might be the basis for an agent's motivating reason for performing it), but also on whether it has any outweighing bad feature. Thus, doing something to protect a dog from a boy's kick is permissible when it involves blocking that kick with one's own leg, but not when it involves shooting him before his kick has a chance to connect. So, to ensure counterfactual reliability while adopting the Simple Humean View, we would have to adopt the fundamentalist version of the Simple Humean View. After all, it's only a nonderivative concern to perform acts that have the *fundamental* right-maker that will ensure that one never does wrong when relevantly informed.

But the fundamentalist version of the Simple Humean View is also unacceptable. The problem is that it makes moral worth too hard to come by. As Daniel Star (Manuscript) has pointed out, people rarely conceptualize

---

[12] I'm assuming that the boy's blow would cause the dog a lot more harm if not blocked than it would cause her if blocked.

their actions as meeting some fundamental moral criterion.[13] In any case, *The Empathic* is a clear counterexample to the fundamentalist version of the Simple Humean View. Christoforos wasn't motivated out a nonderivative concern for anything such as maximizing utility, abiding by the ideal code of rules, or acting in accordance with the categorical imperative. Rather, he was motivated simply out of a nonderivative desire to alleviate what he correctly perceived to be the chimpanzee's distress. What's more, he had all the other pertinent concerns. For he cared about the welfare of both the poachers and the authorities. And he even cared about doing what's right. He just didn't have an additional concern for, say, doing what would maximize utility, conceived as such. But caring about each individual and in the correct proportions (in, say, proportion to the amount of welfare that's at stake for each of them) will unerringly lead him to maximize utility (and, thus, to act rightly) in any situation in which he is relevantly informed. Therefore, we should also reject the fundamentalist version of the Simple Humean View.

4. Where These Two Simple Views Go Wrong

We should reject both the Simple Kantian View and the Simple Humean View. We should reject the Simple Kantian View because it gets the wrong verdict in *The Empathic*. And we should reject the Simple Humean View because it gets the wrong verdict in either *The Empathic* or *The Dog-Lover*, depending on which version we're considering.[14] Now, there have been several attempts to salvage some version of these two views. But I doubt that either can be salvaged, for they both go wrong in a very fundamental way. Specifically, they both go wrong in failing to acknowledge that all and only those right acts that issue from an appropriate set of concerns and the relevant knowledge have moral worth.

More specifically, the Simple Kantian View goes wrong in insisting that acts with moral worth must manifest a nonderivative concern for doing

---

[13] See Star (Manuscript). See also Howard (2021a).

[14] Jessica Isserow has recently defended a pluralist proposal according to which "it is necessary and sufficient for an agent's action to have moral worth that she be motivated either by the consideration that her action is morally right, or by the considerations that explain why her action is morally right" (2020, 550). I believe that her view will suffer the same fate as the Simple Humean View, because either "the considerations that explain why the agent's act is right" will refer exclusively to the act's fundamental right-maker or it won't. If it does, then it will get the wrong result in *The Empathic*. And if it doesn't, then it will get the wrong result in *The Dog-Lover*.

what's right, when, arguably, having a *nonderivative* concern for doing what's right is inappropriate. For it seems that we should not care about doing what's right for its own sake. Rather, we should care about doing what's right only because we care about the ends at which morality ultimately aims and know that doing what's right is a means to our best furthering those ends. Thus, we should care about acting morally, not for its own sake, but only for the sake of the ends at which morality ultimately aims. Nathan Howard makes this point nicely.

> Acting from a desire for rightness as such . . . is a little like desiring to get the cheap plastic trophy without caring about whether you're the champion. The trophy is worth getting only because it represents the verdict that you're the champ. Therefore, desiring the cheap plastic trophy as such fetishizes the trophy; it displaces your desire from its fitting object, namely, the end of being the champ. Likewise for rightness. If rightness is worth caring about, it is only derivatively so, in virtue of its connection to the ends at which morality properly aims like equality, welfare, and the care that we owe to our friends, family, and fellow humans. (2021b, 302–4)

Consider an analogy. Suppose I'm a sperm donor who wants my progeny to flourish. Unfortunately, I don't know how many progenies I have, let alone who they are, for the sperm bank insists on strict anonymity. Now, imagine that I can trust Episteme to tell me what I ought to do according to the means-end-rationality standard, and he tells me that I ought, according to this standard, to wire money into certain numbered accounts. What's more, I know that if I do what I ought, according to this standard, to do, my progeny will flourish. In this case, it seems that, although I should care nonderivatively about my progeny, I should care, only derivatively, about doing what I ought, according to this standard, to do. For doing what I ought, according to this standard, to do is merely a means to ensuring that my progeny flourish. This, I believe, is analogous to the situation in which I want to further the ends at which morality ultimately aims but am uncertain as to what those ends are. In such a case, I should care about doing what I morally ought to do only as a means to furthering the ends that I nonderivatively care about: the ends at which morality ultimately aims, whatever they may be.

The Simple Humean View also goes wrong; it does so in denying that an act can have moral worth in virtue of manifesting a derivative concern for doing what's right, when, arguably, such a concern is entirely appropriate.

For suppose that I care nonderivatively about the ends at which morality ultimately aims and, so, want to do what will best further those ends. But suppose that I don't know how best to further those ends, either because I don't know what they are or don't know how best to further them. I may, nevertheless, know—as a result of, say, reliable testimony—what the right thing to do is. What's more, I may know that I can best further the ends at which morality ultimately aims by doing what's right. And, in such instances, I should have a derivative concern for doing what's right—one that's derivative of my nonderivative concern for the ends at which morality ultimately aims. Indeed, as Sliwa points out, "conative states with moral content (e.g., a desire to do what's right) are essential for doing the right thing in the face of moral uncertainty" (2016, 408).[15] Also, there will be times when one will be tempted to do wrong because one fails in the moment to be appropriately moved by the things that ultimately matter. And, in such cases, being moved to do what's right can serve as a proxy for being moved directly by the things that ultimately matter.[16] Yet, according to the Simple Humean View, an act has moral worth only if it manifests a nonderivative concern for its right-making features, and a derivative concern for an act's being right isn't the same as having a nonderivative concern for its right-making features. So, we must reject the Simple Humean View.

As we've just seen, both the Simple Kantian View and the Simple Humean View fail to accommodate the plausible idea that all and only those right acts that issue from an appropriate set of concerns and the relevant knowledge have moral worth, where the appropriateness of a concern—both in terms of its magnitude and in terms of its being either derivative or nonderivative—is determined by what our ultimate moral concerns should be. The Simple Kantian View fails in requiring us to have an inappropriate concern (specifically, a nonderivative concern for doing what's right), and the Simple Humean View fails in prohibiting us from having an appropriate concern (specifically, a derivative concern for doing what's right). Of course, in making these arguments, I've relied heavily on the notion of what our ultimate moral concerns should be. Unfortunately, this notion has been undertheorized. So, I will need to take a brief digression from our discussion of moral worth to explicate it in the next section.

---

[15] Although she's right about this, she's wrong to assume that "what's important is that the concern for doing what's right be non-instrumental" (Sliwa 2016, 396). She claims that "the agent must care about doing what's right for its own sake, and not because it would further some other goal" (2016, 396).

[16] See Lillehammer (1997) for some nice examples.

5. Our Ultimate Moral Concerns

Unfortunately, many contemporary moral theories fail to tell us what our ultimate moral concerns should be. That is, they fail to tell us what we, as moral agents, should ultimately be aiming to achieve. There are at least three reasons why a complete moral theory owes us an account of this. First, the question of what we, as moral agents, should ultimately be aiming to achieve is itself an important moral question and, thus, one that a complete moral theory should answer for us. Second, we need to know what our ultimate moral concerns should be so that we can determine whether our concerns are appropriate and, in turn, whether the acts that issue from them have moral worth. And, third, whether the ultimate moral concerns that a moral theory prescribes for us is consistent with our being motivated to do as its criterion of rightness directs us to act determines whether it is *incoherent*— incoherent in that its criterion of rightness sometimes permits (or, even worse, requires) agents to act in ways that they know won't achieve the ends that the theory directs them to achieve. And this is important, because we should reject such a theory.

To see why, consider the incoherence objection to rule-consequentialism. According to rule-consequentialism's criterion of rightness, an act is morally permissible if and only if it accords with the ideal code of rules. Now, some have worried that the ideal code will be extensionally equivalent to act-consequentialism and that, therefore, rule-consequentialism will collapse into act-consequentialism. But, as Brad Hooker (2000) has shown, rule-consequentialism can avoid collapsing into act-consequentialism. But, in avoiding the collapse worry, a new worry arises. For if the ideal code isn't extensionally equivalent to act-consequentialism, then there will be instances in which rule-consequentialism permits (or even requires) an agent to abide by the ideal code even though they know that their doing so won't maximize the good. And, so, if a complete version of rule-consequentialism holds both that agents must adopt maximizing the good as their ultimate moral concern and that agents are sometimes permitted (or even required) to abide by the ideal code even when they know that doing so won't maximize the good, then it will be an incoherent theory. That is, it will require them to have incoherent motives. On the one hand, they'll be required to abide by (and to be motivated to abide by) the ideal code even when they know that doing so won't maximize the good. And, on the other hand, they'll be required to adopt maximizing the good as their *ultimate* moral concern, such that they will be concerned with abiding by the ideal code only as a means to maximizing the good. But if they're concerned with abiding by the ideal code only as a means to maximizing the good, then they won't be motivated, as required, to abide by the ideal code even when they know that doing so won't maximize the good.

Hooker's response to the objection is to deny that a complete version of rule-consequentialism must give each agent the ultimate moral concern of maximizing the good. He says, "rule-consequentialists need not have maximizing the good as their ultimate moral goal" (2000, 101). He holds that rule-consequentialism is itself committed only to both a certain conception of the good (which tells us how to assess the goodness of various codes of rules) and a certain conception of the right (which tells us how to assess the rightness of acts in terms of the goodness of the various codes of rules that either permit or prohibit them), but not to any particular conception of what our ultimate moral concerns should be. So, Hooker, qua rule-consequentialist, can deny that agents should have maximizing the good as their ultimate moral concern and hold instead that they should have ensuring that their acts are impartially defensible as their ultimate moral concern.[17] Indeed, this is what he does. And this allows him to avoid the incoherence objection, because there is nothing incoherent about a theory that holds both that agents should have acting only in ways that are impartially defensible as their ultimate moral concern and that agents should abide by (and be motivated to abide by) the ideal code even when they know that doing so won't maximize the good. For such a theorist can just claim that acting in accord with the ideal code ensures that one's acts are impartially defensible even when those acts fail to maximize the good.

So, it's crucial that a complete moral theory tell us what our ultimate moral concerns should be so that we can then determine whether it's coherent. Now, admittedly, many contemporary moral philosophers have ignored this aspect of moral theorizing. They have often contented themselves with merely offering a criterion of rightness. Or if they go beyond that, they do so only to include a decision procedure. There are exceptions, of course. Nevertheless, contemporary moral philosophers have tended to neglect the issue of what our ultimate moral concerns should be. But this is clearly a mistake given that it's not just rule-consequentialism that's

---

[17] Perhaps, even this isn't what our ultimate moral concern should be. For perhaps we should be concerned with our acts being impartially defensible only because we should be concerned to show respect for people's humanity, which requires ensuring that we act only in ways that are impartially defensible.

potentially subject to an incoherence objection. All moral theories are potentially subject to this objection. Take, for instance, maximizing act-utilitarianism. It will be incoherent if it holds that agents should have an ultimate moral concern for ensuring that each sentient creature has as much utility as possible. To see why, note both that (1) maximizing act-utilitarianism holds that an act is permissible if and only if there is no alternative act that would produce a greater sum of utility than it would and that (2) although some infinities are, in some sense, larger than others, the sum of a denumerably infinite number of locations with 2 hedons each is not greater than the sum of an equal number of locations with 1 hedon each (see Kagan and Vallentyne 1997). Given these assumptions, maximizing act-utilitarianism implies that you would be permitted to φ and thereby provide an infinite number of sentient creatures with 1 hedon each even if you could instead have ψ-ed and thereby provided them with 2 hedons each. Such a theory would be incoherent, because it permits you to act in a way that you know won't optimally advance the ultimate moral aim that it gives you. For it permits you to φ even though you know that φ-ing won't optimally advance the ultimate moral concern that you have for ensuring that each sentient creature has as much utility as possible.

We've seen, then, that the notion of an ultimate moral concern is crucial to moral theorizing, for we should reject theories that are incoherent, and whether a theory is incoherent just depends on its account of what our ultimate moral concerns should be. We've also seen that such an account is crucial to determining whether the concerns that issue in an act are appropriate, which, in turn, is relevant to determining its moral worth. Thus, in the next section, I return to the issue of moral worth and show how this notion of an ultimate moral concern can help us to develop a plausible alternative to both the Simple Kantian View and the Simple Humean View.

6. The Concerns View and How It Compares to Its Rivals

I can now state what I take to be the correct account of moral worth.

The Concerns View: A right act has moral worth if and only if it issues from an appropriate set of concerns and the knowledge (or, at least, the rational belief) that this is what would best further those concerns, where this set includes all and only pertinent concerns, each of which must be both qualitatively and quantitively appropriate, which in turn depends on what the agent's ultimate moral concerns should be.[18]

Remember that the pertinent concerns are all and only those that will (or would) determine whether the agent acts rightly in this and other relevantly similar situations—the relevantly similar situations being determined by the context. And whether a given concern is both qualitatively and quantitively appropriate depends on what the agent's ultimate moral concerns should be. To illustrate, suppose, as I argued above, that acting rightly is not something for which an agent should have an ultimate moral concern. In that case, they should be concerned with acting rightly only insofar as acting rightly is a means to furthering the ends for which they should have an ultimate moral concern. And, thus, their concern for acting rightly will be qualitatively appropriate only if it's derivative in this way. What's more, it will be quantitatively appropriate only if its magnitude is in direct proportion to the extent to which their acting rightly is, in the given situation, a means to their furthering the ends for which they should have an ultimate moral concern.

To take another example, imagine that we should have an ultimate moral concern for promoting each existing individual's utility but not for promoting the overall sum of utility in the universe. In that case, it would be inappropriate for us to want to bring happy individuals into existence for the sake of increasing the overall sum of utility, but appropriate for us to want to do so as a means to promoting the utility of existing individuals, as where they would derive utility from our bringing these happy individuals into existence. Let's further assume that, given the appropriate ultimate moral concerns, we should care just as much about n hedons of utility for one stranger as we do about n hedons of utility for any other stranger. Thus, how intensely we should want to promote the utility of an existing stranger by n hedons should depends solely on how great the number n is.

Now that we have a sense of how an agent's ultimate moral concerns determine the appropriateness of the concerns from which their act stems, we can look at how the Concerns View deals with various cases, starting with *The Dog-Lover*. On the Concerns View, Yunn's act lacks moral worth given that she lacks a pertinent concern: specifically, a concern for the welfare of

---

[18] I've added an epistemic condition here, because it seems that even if one has the appropriate set of concerns, one's act will count as merely accidentally right if one performs this act only because one irrationally (although, by luck, correctly) believes that this act is what will best further those concerns. I thank Ralph Wedgwood for pointing this out.

the boy. For it's plausible to suppose that, for each sentient being, Yunn should have an ultimate moral concern for promoting that being's welfare. Yet, she has zero concern for the boy's welfare. What's more, this concern is a pertinent one given that it would in combination with other appropriate concerns lead her to act wrongly in various relevantly similar situations, such as the one in which she has the option of shooting the boy. And since Yunn's act of blocking the boy's blow doesn't stem from an appropriate set of concerns, the Concerns View rightly implies that it lacks moral worth. And there's a lesson here.

Lesson 1:   A right act can lack moral worth even if it was performed for the right reasons—that is, for the reasons that make it right.

After all, Yunn did perform the right act, and she did so for the right reason (i.e., to protect the dog). Nevertheless, her act lacks moral worth, for it was merely fortuitous that she did the right thing. Had the situation been slightly different, she would have been led by the same set of concerns to act wrongly. And, thus, we should reject those rivals of the Concerns View that insist that moral worth is a matter of acting for the right reasons—views such as those defended by Markovits (2010, 205).

   Turn now to *The Empathic*. On the Concerns View, it's not only Christoforos's act of notifying the authorities but also his act of helping the chimpanzee escape that has moral worth. Both have moral worth because both stem from an appropriate set of concerns and the relevant knowledge. In notifying the authorities and hiding the poacher's guns, Christoforos manifests a concern (a derivative concern) for doing what's right as well as a concern (a nonderivative concern) for promoting the welfare of each of the sentient beings involved. And these concerns are all quantitatively appropriate—or so we're assuming. Likewise, his act of helping the chimpanzee escape stems from an appropriate set of concerns. Of course, it doesn't manifest a nonderivative concern for doing what's right given that he doesn't think that helping the chimpanzee escape is right; he thinks, rather, that it's morally neutral. Nevertheless, his act does manifest an empathetic concern for alleviating what he correctly perceives to be the chimpanzee's distress. And this concern is entirely appropriate given that he should have an ultimate moral concern for the chimpanzee's welfare and, thus, derivatively, for doing what would alleviate its distress. So, in this case too, the Concerns View gets the intuitive verdict. What's more, the Concerns Views provides us with another important lesson.

Lesson 2: An act can have moral worth even if it doesn't manifest a nonderivative concern for doing what's right.

Thus, we should reject those rivals of the Concerns View that insist that, for an act to have moral worth, it must manifest a nonderivative concern for doing what's right—views such as those defended by Herman (1981), Sliwa (2016), Johnson King (2020), and Singh (2020).

The Concerns View not only allows that an act can have moral worth without manifesting a *nonderivative* concern for doing what's right but also allows that an act can have moral worth in virtue of manifesting a *derivative* concern for doing what's right. And this is important, because, as many have pointed out, such a concern helps agents deal with both temptation and uncertainty. Sometimes, we're tempted to do what's wrong but are moved by our derivative desire to do what's right to resist that temptation. Other times, we do not know (or are uncertain about) what we should ultimately care about. But if we know what it's right to do and that by doing what's right we can best further the ends that we should ultimately care about, then we can do what's right as a means to furthering the ends that we should ultimately care about. So, although there is something problematic about being motivated out of a *nonderivative* concern to do what's right given that this is fetishistic (see Smith 1994), there's nothing problematic about being motivated by a *derivative* concern to do what's right where one faces temptation or uncertainty. And, so, we derive yet another important lesson.

Lesson 3: An act can have moral worth in virtue of its manifesting a concern (specifically, a derivative concern) for doing what's right.

And this means that we should reject rivals of the Concerns View that insist that an act can't have moral worth in virtue of its manifesting a concern for doing what's right—views such as those defended by Arpaly (2002).

Lastly, cases such as *Golfing for Rightness' Sake* show us that being motivated to do what's right because it's right is insufficient to confer moral worth on an act. Consider that, in this case, each doctor takes her husband golfing because it's the right thing to do, and, yet, their acts are only accidentally right given that they would have done the exact same thing even if it had been wrong to do so. So, deliberately acting rightly is insufficient to confer moral worth on an act. And this brings us to our fourth and final lesson.

Lesson 4: A right act can lack moral worth even if its agent performs it because it's right.

So, we should reject rivals of the Concerns View that insist that an act will have moral worth if it is an instance of its agent deliberately acting rightly—views such as those defended by Johnson King (2020).

We've seen, then, that the Concerns View is superior to its rivals. Unlike the Simple Humean View and its contemporary descendants (e.g., Markovits 2010), it allows that a right act can lack moral worth even if it was performed for the right reasons. And, unlike the Simple Kantian View and its contemporary descendants (e.g., Johnson King 2020), it holds that an act can have moral worth even if it doesn't manifest a concern for doing what's right.[19]

References

Arpaly, N. (2002). "Moral Worth." *Journal of Philosophy* 99: 223–45.

Arpaly, N. (2015). "Moral Worth and Normative Ethics." *Oxford Studies in Normative Ethics Vol. 5*: 86–105. Oxford: Oxford University Press.

Arpaly, N., and T. Schroeder (2014). *In Praise of Desire*. Oxford: Oxford University Press.

Baron, M. (1995). *Kantian Ethics Almost without Apology*. Ithaca, NY: Cornell University Press.

Estlund, D. (2017). "Prime justice." In *Political Utopias*, K. Vallier and M. Weber (eds.), 35–55. Oxford: Oxford University Press.

Furtak, R. A. (2018). *Knowing Emotions*. Oxford: Oxford University Press.

Hanser, M. (2005). "Permissibility and Practical Inference." *Ethics* 115: 443–70.

Herman, B. (1981). "On the Value of Acting from the Motive of Duty." *Philosophical Review* 90: 359–82.

Hooker, B. (2000). *Ideal Code, Real World*. Oxford: Oxford University Press.

Howard, N. R. (2021a). "The Goals of Moral Worth." In *Oxford Studies in Metaethics Vol. 16*: 157–82. Oxford: Oxford University Press.

---

[19] For helpful comments and discussions, I thank Ron Aboodi, Chrisoula Andreou, Cheshire Calhoun, Jamie Dreier, Josh Glasgow, Peter Graham, Brad Hooker, Nathan Howard, Paul Hurley, Jessica Isserow, Zoë Johnson King, Andrew Khoury, Sarah McGrath, Aaron Rizzieri, Grant Rozeboom, Lucia Schwarz, David Shoemaker, Keshav Singh, Holly Smith, Hannah Tierney, Travis Timmerman, Mark Timmons, Ralph Wedgwood, Erik Zhang, and an anonymous reviewer.

Howard, N. R. (2021b). "One Desire Too Many." *Philosophy and Phenomenological Research* 102: 302–17.

Hume, D. (2007). *A Treatise of Human Nature: A Critical Edition*. D. F. Norton and M. J. Norton (eds.). Oxford: Clarendon Press. [Originally published in 1739–40.]

Hurka, T. (2006). "Virtuous Act, Virtuous Dispositions." *Analysis* 66: 69–76.

Isserow, J. (2020). "Moral Worth: Having it Both Ways." *Journal of Philosophy* 117: 529–56.

Johnson King, Z. A. (2020). "Accidentally Doing the Right Thing." *Philosophy and Phenomenological Research* 100: 186–206.

Kagan, S., and P. Vallentyne (1997). "Infinite Value and Finitely Additive Value Theory." *Journal of Philosophy* 94: 5–26.

Kant, I. (1998). *Groundwork of the Metaphysics of Morals*. Trans. M. Gregor. Cambridge: Cambridge University Press. [Originally published in 1785.]

Lillehammer, H. (1997). "Smith on Moral Fetishism." *Analysis* 57: 187–95.

Markovits, J. (2010). "Acting for the Right Reasons." *Philosophical Review* 119: 201–42.

Singh, K. (2020). "Moral Worth, Credit, and Non-Accidentality." *Oxford Studies in Normative Ethics Vol. 10*: 156–81. Oxford: Oxford University Press.

Sliwa, P. (2016). "Moral Worth and Moral Knowledge." *Philosophy and Phenomenological Research* 93: 393–418.

Smith, M. (1994). *The Moral Problem*. Oxford: Blackwell.

Star, D. (Manuscript). "Moral Worth, Normative Ethics, and Moral Ignorance."

Index

For the benefit of digital users, indexed terms that span two pages (e.g., 52–53) may, on occasion, appear on only one of those pages.